T0239062

SpringerBriefs in Applied Sciences and Technology

Computational Mechanics

Series Editors

Holm Altenbach⬡, Faculty of Mechanical Engineering,
Otto-von-Guericke-Universität Magdeburg, Magdeburg, Sachsen-Anhalt, Germany

Lucas F. M. da Silva, Department of Mechanical Engineering, Faculty of
Engineering, University of Porto, Porto, Portugal

Andreas Öchsner, Faculty of Mechanical Engineering, Esslingen University of
Applied Sciences, Esslingen, Germany

These SpringerBriefs publish concise summaries of cutting-edge research and practical applications on any subject of computational fluid dynamics, computational solid and structural mechanics, as well as multiphysics.

SpringerBriefs in Computational Mechanics are devoted to the publication of fundamentals and applications within the different classical engineering disciplines as well as in interdisciplinary fields that recently emerged between these areas.

More information about this subseries at http://www.springer.com/series/8886

Arshad Afzal · Kwang-Yong Kim

Analysis and Design Optimization of Micromixers

 Springer

Arshad Afzal
Mechanical Engineering
Indian Institute of Technology Kanpur
Kanpur, India

Kwang-Yong Kim
Mechanical Engineering
Inha University
Incheon, Korea (Republic of)

ISSN 2191-530X   ISSN 2191-5318   (electronic)
SpringerBriefs in Applied Sciences and Technology
ISSN 2191-5342   ISSN 2191-5350   (electronic)
SpringerBriefs in Computational Mechanics
ISBN 978-981-33-4290-3   ISBN 978-981-33-4291-0   (eBook)
https://doi.org/10.1007/978-981-33-4291-0

This Springer imprint is published by the registered company Springer Nature Singapore Pte Ltd.
The registered company address is: 152 Beach Road, #21-01/04 Gateway East, Singapore 189721, Singapore

Contents

Abbreviations

3D	Three-dimensional
CFD	Computational fluid dynamics
CV	Cross-validation
DOE	Design of experiment
GA	Genetic algorithm
KRG	Kriging
LHS	Latin hypercube sampling
MLE	Maximum likelihood estimation
MOGA	Multi-objective genetic algorithm
NSGA	Non-dominated sorting genetic algorithm
PDMS	Polydimethylsiloxane
PMMA	Poly-methyl methacrylate
POD	Pareto-optimal design
PSO	Particle swarm optimization
RBNN	Radial basis neural networks
RSA	Response surface approximation
SA	Simulated annealing
SAR	Split and recombine
SHM	Staggered herringbone micromixer
SQP	Sequential quadratic programming
TLCCM	Two-layer crossing channel micromixers
μ-TAS	Micro-total analysis system

Nomenclature

c	Neurons centers
C	Species concentration
C_m	Optimal concentration
d	Input dimension
D	Diffusion coefficient
f	Objective function
g	Regressors in Kriging model
k	Number of clusters
K	Number of neurons centers
L	Characteristic length scale
M	Mixing index
n	Number of points on cross-sectional plane
N	Sample size
p	Dimensional pressure
Pe	Peclet number
R	Correlation function
R^2_{adj}	Goodness of fit
Re	Reynolds number
Sc	Schmidt number
t	Dimensional time
U	Characteristic velocity scale
U_i	Components of velocity vector (i = 1, 2 and 3)
V	Fluid velocity
V_p	Particle velocity
x_i	Dimensional co-ordinates (i = 1, 2 and 3)
x_p	Particle position
X	Design matrix
$X^\dagger$	Pseudo-inverse of X

$\bar{y}$	Trend function in Kriging model
Y	Vector of response
Z	Random field with zero mean and stationary covariance

Greek Symbols

α	Regression coefficient in Kriging model
β	Regression coefficient
γ	Parameter in radial basis function
$\in$	Error
θ	Parameter of Gaussian kernel in Kriging model
μ	Dynamic viscosity of fluid
v	Kinematic viscosity of fluid
ρ	Density of fluid
σ	Variance of concentration
$\sigma_{\in}$	Error variance
σ_f	Process variance
ϕ	Gaussian function for radial basis neuron
φ	Polynomial basis functions

Chapter 1
Mixing at Microscale

Abstract Micromixers are essential components of lab-on-a-chip and micro-total analysis systems used for a variety of chemical and biological applications such as sample preparation and analysis, protein folding, DNA analysis, and cell separation. Due to the small characteristic dimension of micromixers, the flow is laminar in a Reynolds number range from 0.01 to 100 for typical microfluidic applications. In microfluidic devices, the laminar flow condition poses a challenge for the mixing of liquid samples. Therefore, for high performance lab-on-a-chip and micro-total analysis systems, it is essential to develop and devise micromixers to achieve fast and compact mixing at the micro-scale. Although mixing can involve different phases (solid, liquid and gases), the present book focuses on liquid–liquid mixing such as water–ethanol mixing. This chapter provides an introduction to application of micromixers, flow dynamics and mixing in micromixers, and dimensionless numbers which characterize flow and mixing regimes.

Keywords Mixing at microscale · Micromixer · Laminar flow · Lab-on-a-chip · Micro-total analysis systems

1.1 Applications of Micromixers

Microfluidic systems, such as lab-on-a-chip and micro-total analysis system (μ-TAS) with dimensions of the order of microns have gained widespread attention. Micromixers have emerged as an important component of lab-on-a-chip and μ-TAS used for wide variety of chemical and biological applications [1–9]. A micromixer aims to mix two or more samples produced from prior processes of a microfluidic system. The chemical applications of micromixers include chemical synthesis, polymerization, and extraction, while the biological applications include DNA analysis, biological screening enzyme assays and protein folding. Also, micromixers are main component of microreactors, and used during the manipulation of reagents and catalyst concentration. The majority of applications required Reynolds number (Re) in a range of $0.1 \leq \mathrm{Re} \leq 100$.

© The Author(s), under exclusive license to Springer Nature Singapore Pte Ltd. 2021   1
A. Afzal and K.-Y. Kim, *Analysis and Design Optimization of Micromixers*,
SpringerBriefs in Computational Mechanics,
https://doi.org/10.1007/978-981-33-4291-0_1

Ottino and Wiggins [3] provided two distinct applications of mixing. The first one is related to protein folding (a process by which proteins assume their unique three-dimensional (3D) shapes) in which molecular diffusion solely controls the mixing. The second one is rapid mixing of macromolecular solutions for chip-based molecular diagnostics (immunoassays and hybridization analyses, which require the rapid, homogeneous mixing of macromolecular solutions, such as DNA and globular proteins), and it involves intermixing of two streams. For the latter application, mixing by pure molecular diffusion will take considerable time, and it becomes important to find suitable micromixer designs to mix fluid streams in a short time over a short mixing length.

Li et al. [4] demonstrated a unique continuous-flow laminar mixer based on microfluidic dual-hydrodynamic focusing to characterize the kinetics of DNA–protein interaction. The novel micromixer was also found useful for analyzing the interaction kinetics of biomacromolecules. Hessel et al. [5] used the IMM (Institut für Mikrotechnik Mainz GmbH) liquid split-and-recombine (SAR) passive micromixer for the aqueous Kolbe-Schmitt synthesis using resorcinol to yield 2, 4-dihydroxy benzoic acid. Anwar et al. [6] successfully demonstrated and applied a passive micromixer based on unbalanced split and cross collisions of fluid streams to sample preparation and preconcentration of proteins on a microfluidic platform for biosensor applications.

Stone and Kim [7] provided a brief review on basic issues, applications and challenges encountered in the design and development of microfluidic devices. Some interesting aspects of the research were the importance of scaling down devices, fabrication techniques, and the effects of driving forces, such as pressure difference, electric fields and surface tension, on fluid flow in a microchannel. The paper also highlighted the interdisciplinary nature of researches in microfluidics with different branches of science and engineering come together to realize devices with specific functions. Jeong et al. [10] presented a large number of practical applications of different types of passive micromixers. Micromixers in various applications were grouped according to Reynolds number and mixing performance.

Lee and Fu [11] presented a comprehensive review on state-of-the-art biomedical applications of micromixers. Various micromixers were classified based on their use, particularly for sample concentration, chemical synthesis and reaction, polymerization, extraction and purification, biological analysis, and droplet/emulsion process. A statistical analysis of the published researches from 2004 to 2017 in terms of the application of micromixers revealed that the main area is chemical reactor followed by biological and chemical analyses. The review provided a nice understanding of micromixer designs and their related applications.

1.2 Micromixer Types

The characteristic dimension of micromixers is in the sub-millimeter range. Typical channel width and height of micromixers are in the range of 100–500 μm, while the channel length can be up to few millimeters (limited by the dimension of the microfluidic platform for a particular application). Compared with large-scale mixing devices, microfluidic systems offer advantages in terms of lower sample consumption, a cheaper manufacturing cost, and higher throughput.

Mixing at microscale is challenging due to laminar flow ($0.01 \leq Re \leq 100$) inside the channel on account of dominant viscous force (Fig. 1.1). Therefore, the random turbulent fluctuations necessary to homogenize fluid samples are absent at the microscale. Thus, the main transport phenomena in micromixers are limited to diffusion and advection. Diffusive mixing in laminar flows (Fig. 1.2) is considerably slow, and requires long channel lengths for complete mixing, but the channel length is limited by the size of the microfluidic platform. Diffusion mechanism is characterized by a molecular diffusion coefficient (typical value for the diffusivity of small proteins in an aqueous solution is 10^{-10} m^2 s^{-1}). Advection causes stretching and folding of fluid interface providing distinct advantages. It leads to a greater interfacial area which means a greater area for mass transfer, and at the same time, reduction in the striation thickness increases the concentration gradient and accelerates the mixing process.

To enhance mixing at microscale, many researchers have developed different types of micromixers, which can be broadly classified into two types: active and passive micromixers. Figures 1.3 and 1.4 show some selected designs of passive [12–15] and active [16–18] micromixers, respectively. Passive micromixers rely on the geometry of the flow passage to produce complex flow fields for effective mixing of the fluid samples, and do not require any external energy. On the other hand, the active type uses moving parts or some external agitation/energy for the mixing. Magnetic energy,

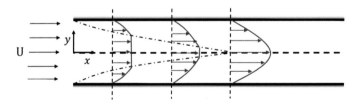

Fig. 1.1 Laminar flow in a channel

Fig. 1.2 Diffusive mixing in laminar flow

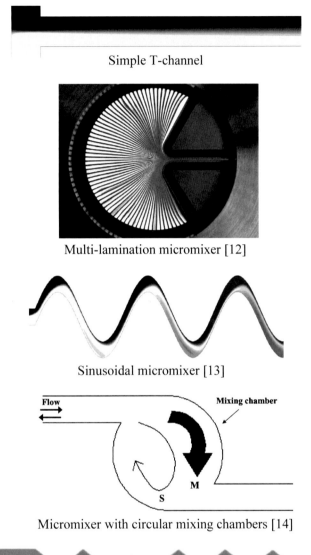

Simple T-channel

Multi-lamination micromixer [12]

Sinusoidal micromixer [13]

Micromixer with circular mixing chambers [14]

Two-layer crossing channel micromixer [15]

Fig. 1.3 Passive micromixers

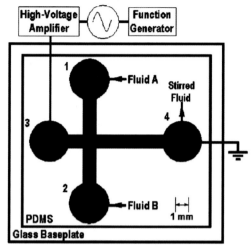

Electrokinetic instability micromixer [16]

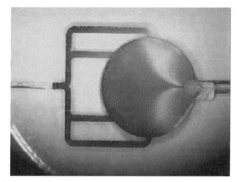

Micromixer based on Acoustic streaming [17]

Pulsed-flow micromixer [18]

Fig. 1.4 Active micromixers

electrical energy, pressure disturbance, and ultrasonic mixing can be used to stir the fluids. Therefore, active micromixers provide higher mixing performance, while passive micromixers offer the advantage of simple design and fabrication, and can be easily integrated to a complex system.

Hardt et al. [12] categorized passive micromixers based on the hydrodynamic principle employed, viz. chaotic advection, flow separation, hydrodynamic focusing, and split-and-recombination of flows. The review aimed to provide the micromixer designers to select the most favorable concept for a particular application. The review by Nguyen and Wu [19] reported the development of micromixers based on different working principles. Both active and passive types were discussed, but the major emphasis was laid on the design and mechanism of passive micromixers. A brief discussion on operating conditions, fabrication techniques and mixing characterization was also included in the review.

Hessel et al. [20] conducted a detailed review on passive and active mixing principles and described the typical mixing element designs, methods for mixing characterization, and application fields. The review also discussed the mixing of gases in microchannels. Kumar et al. [21] reported the operating ranges of passive and active micromixers in terms of Reynolds and Peclet numbers using an extensive literature survey, as shown in Fig. 1.5. Compared to the active counterpart, passive micromixers were reported for wider ranges of Reynolds ($0.001 \leq Re \leq 1000$) and Peclet ($0.01 \leq Pe \leq 1000, 000$) numbers. Micromixers designed for diffusive mixing work well only at low Reynolds and Peclet numbers, but micromixers designed for chaotic advection can be used for a wide range of Reynolds numbers.

1.3   Mechanism of Mixing

Micromixers can operate over a wide Reynolds number range [20, 21]. The mixing mechanism and performance strongly depend on Reynolds number regime. Figure 1.6 shows the variations of mixing index with Reynolds number for three micromixers with different planar serpentine channels: square-wave, zig-zag and curved channels. For low Reynolds numbers ($Re < 1$), mixing occurs primarily by diffusion, which is dominated by the residence time of the co-flowing fluids in the micromixer. As Reynolds number increases, the residence time of the fluids in the mixer decreases, thereby reducing mixing index. After the mixing index reaches a minimum around $Re = 10$, transverse flow starts to occur in the channel and the mixing index increases rapidly with further increase in Reynolds number. This rapid increase in mixing index is attributed to enlargement of fluid interface caused by stretching and folding of fluid layers due to chaotic advection. The concentration contours at $Re = 10$ show that the concentration layers are sharp and well aligned in the microchannel, and the mass transfer takes place by diffusion at the interface between the fluid layers of different concentrations. However, at $Re = 50$, the layers in the channel no longer remain aligned due to stretching and folding of the fluid interface.

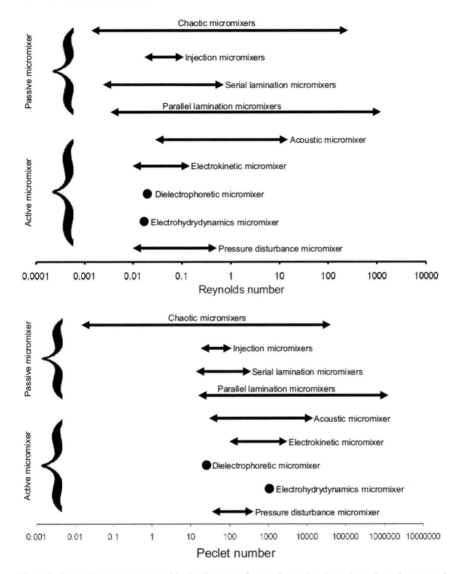

Fig. 1.5 Operating ranges reported in the literature for passive and active micromixers in terms of Reynolds and Peclet numbers [21]

1.4 Dimensionless Parameters

The dynamics of flow through micromixers is governed by the following dimensionless parameters:

1. Reynolds number $(Re) \equiv \frac{UL}{\nu}$
2. Peclet number $(Pe) \equiv \frac{UL}{D}$

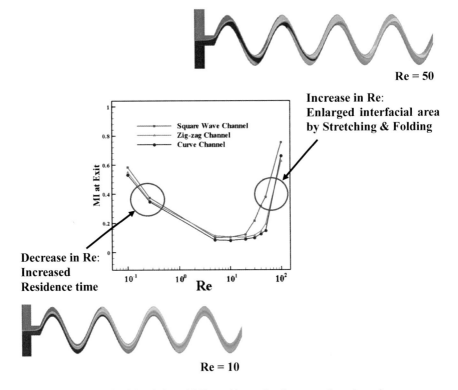

Fig. 1.6 Variations of mixing index with Reynolds number for serpentine micromixers

3. Schmidt number (Sc) $\equiv \frac{\nu}{D}$

where L and U are the characteristic length and velocity scales of the micromixer, respectively. The kinematic viscosity and coefficient of mass diffusivity are denoted by ν and D, respectively.

The Reynolds number represents the ratio of inertial to viscous force. In most microfluidics applications, due to low Reynolds numbers, laminar flow is expected. The Peclet number represents the relative importance of the mass transfer by convection compared to that of diffusion. Convection is dominated at high Peclet numbers. For water flows at room temperature, the kinematic viscosity ν and diffusion coefficient D are 10^{-6} m^2 s^{-1} and 10^{-10} m^2 s^{-1}, which correspond to the Reynolds and Peclet numbers of 0.1 and 1000, respectively, for the cross-sectional dimension, L $= 100$ μm and velocity, U $= 1$ mm s^{-1}. The Schmidt number represents the ratio of momentum to mass transport, and thus the ratio of Peclet to Reynolds number.

1.5 General Comments

The literature survey reveals that there has been a large number of researches on mixing in micromixers from mathematical, theoretical and application perspectives. Various successful micromixers have been reported for enhanced mixing performance over wide ranges of operating conditions. However, there is still room for further improvement in the mixing performance.

Computational fluid dynamics (CFD) has been proved to be reliable for both qualitative and quantitative analyses of species concentration, flow structures, and mixing behavior in micromixers. Preliminary evaluation of micromixer designs can be performed by a parametric study using numerical analysis, which finds the design parameters to which the mixing performance is sensitive. A suitable optimization strategy can be employed to maximize or minimize performance measures, and determine robust and efficient designs of micromixers. In this book, the primary focus is on the application of CFD and optimization techniques to the analysis, design and development of micromixers.

This book is organized as follows: In Chap. 2, both active and passive micromixers are introduced with suitable examples, focusing mainly on design and development of passive micromixers. Chapter 3 deals with the numerical formulation for the analyses of flow and mixing in micromixers, and method for quantification of mixing is introduced. The mixing problem is looked from both Eulerian and Lagrangian approaches; the important features are discussed with appropriate references for the understanding of readers. Various approaches to the design of micromixers based on CFD and optimization techniques are introduced in Chap. 4. Among the approaches, special importance is given to the surrogate-based optimization framework, and a detailed discussion on various facets of the optimization problem is presented. Finally, Chap. 5 summarizes the details of what have been covered in the previous chapters, and suggests how to pursue the researches on micromixers further.

References

1. Service RF (2009) Miniaturization puts chemical plants where you want them. Science 282:400. https://doi.org/10.1126/science.282.5388.400
2. Gambin Y, VanDelinder V, Ferreon ACM, Lamke EA, Groisman A, Deniz AA (2011) Visualizing one-way protein encounter complex by ultrafast single-molecule mixing. Nature Methods 8: 239–241. https://doi.org/10.1038/nmeth.1568
3. Ottino JM, Wiggins S (2004) Introduction: mixing in microfluidics. Phil Trans R Soc Lond 362:923–935. https://doi.org/10.1098/rsta.2003.1355
4. Li Y, Xu F, Liu C, Xu Y, Feng X, Liu BF (2013) A novel microfluidic mixer based on dual-hydrodynamic focusing for interrogating the kinetics of DNA-protein interaction. Analyst 138:4475–4482. https://doi.org/10.1039/c3an00521f
5. Hessel V, Hofmann C, Lob P, Lohndorf J, Lowe H, Ziogas A (2005) Aqueous Kolbe-Schmitt synthesis using resorcinol in a microreactor laboratory rig under high-p, T conditions. Org process Res Div 9:479–489. https://doi.org/10.1021/op050045q

6. Anwar K, Han T, Yu S, Kim SK (2010) An integrated micro-nanofluidic system for sample preparation and preconcentration of proteins. 14th International conference on Miniaturized Systems for Chemistry and Life Sciences, Groningen, The Netherlands

7. Stone HA, Kim S (2001) Microfluidics: Basic issues, applications and challenges. A.I.Ch.E. Journal 47:1250–1254. https://doi.org/10.1002/aic.690470602

8. Meldrum DR, Holl MR (2002) Microfluidics: Microscale bioanalytical systems. Science 297:1197–1198. https://doi.org/10.1126/science.297.5584.1197

9. Ehrlich DJ, Matsudaira P (1999) Microfluidic devices for DNA analysis. Nanotechnology 17: 315-319. https://doi.org/10.1016/S0167-7799(99)01310-4

10. Jeong GS, Chung S, Kim CB, Lee SH (2010) Applications of micromixing technology. Analyst 135:460–473. https://doi.org/10.1039/b921430e

11. Lee CY, Fu LM (2018) Recent advances and applications of micromixers. Sens Actuat B Chem 259:677–702. https://doi.org/10.1016/j.snb.2017.12.034

12. Hardt S, Drese KS, Hessel V, Schönfeld F (2005) Passive micromixer for applications in the microreactor and µTAS fields. Microfluid Nanofluid 1:108–118. https://doi.org/10.1007/s10 404-004-0029-0

13. Afzal A, Kim KY (2013) Mixing Performance of a passive micromixer with sinusoidal channel walls. J Chem Eng Jpn 46:230–238. https://doi.org/10.1252/jcej.12we144

14. Chung YC, Hsu YL, Jen CP, Lu MC, Lin YC (2004) Design of passive mixers utilizing microfluidic self-circulation in the mixing chamber. Lab Chip 4:70–77. https://doi.org/10.1039/B31 0848C

15. Xia HM, Wan SYM, Shu C, Chew YT (2005) Chaotic micromixers using two-layer crossing channels to exhibit fast mixing at low Reynolds numbers. Lab Chip 5:748–755. https://doi.org/10.1039/b502031j

16. Oddy MH, Santiago JG, Mikkelsen JC (2001) Electrokinetic Instability Micromixing. Anal Chem 73:5822–5832. https://doi.org/10.1021/ac0155411

17. Liu RH, Yang J, Pindera MZ, Athavale M, Grodzinski P (2002) Bubble induced acoustic micromixing. Lab Chip 2:151–157. https://doi.org/10.1039/B201952C

18. Afzal A, Kim KY (2015) Convergent-divergent micromixer coupled with pulsatile flow. Sens Actuat B Chem 211:198–205. https://doi.org/10.1016/j.snb.2015.01.062

19. Nguyen NT, Wu Z (2005) Micromixers-a review. J Micromech Microeng 15: R1–R16. https://doi.org/10.1088/0960-1317/15/2/R01

20. Hessel V, Lowe H, Schönfeld F (2005) Micromixers—a review on passive and active mixing principles. Chem Eng Sci 60: 2479–2501. https://doi.org/10.1016/j.ces.2004.11.033

21. Kumar V, Paraschivoiu M, Nigam KDP (2011) Single-phase fluid flow and mixing in microchannels. Chem Eng Sci 66:1329–1373. https://doi.org/10.1016/j.ces.2010.08.016

Chapter 2
Active and Passive Micromixers

Abstract Micromixers are classified into two types: active and passive micromixers. Active micromixers promote mixing using moving parts or some external agitation/energy to stir the fluids. Magnetic energy, electrical energy, pressure disturbance, and ultrasonic are examples of the external energies to enhance mixing. Passive micromixers use geometrical modification to cause chaotic advection or lamination to promote the mixing of the fluid samples, and allow easy fabrication and integration with lab-on-a-chip and μ-TAS. In this chapter, both active and passive micromixers are discussed, but the major emphasis is laid on passive micromixer designs and mechanisms. Extensive referencing on active and passive micromixers is not possible due to the limited length of the book, but the diversity of micromixers is introduced as much as possible.

Keywords Active micromixers · Electrokinetics · Time pulsing · Passive micromixers · Multi-lamination · Chaotic advection

2.1 Active Micromixers

Active micromixers promote mixing using external energy sources such as pressure disturbance, acoustics, vibrations, impellers, and electric and magnetic fields. Based on the type of external energy, the flows in active micromixers can be categorized as:

- Pulsed flow
- Electro-kinetic flow
- Magneto-hydrodynamic flow
- Acoustic induced flow
- Coriolis force induced flow

Glasgow and Aubry [1, 2] conducted a series of extensive studies on pulsed flows by pressure disturbance through T-mixers with different configurations and operating parameters using CFD and experiments. The pulsing was performed using a low frequency sinusoidal flow superimposed upon a steady flow. In one of these studies [2], they investigated the mixing behaviors in ribbed and 3D twisted micromixers

© The Author(s), under exclusive license to Springer Nature Singapore Pte Ltd. 2021 11
A. Afzal and K.-Y. Kim, *Analysis and Design Optimization of Micromixers*,
SpringerBriefs in Computational Mechanics,
https://doi.org/10.1007/978-981-33-4291-0_2

under pulsed flow conditions. The results indicated that mixing was significantly improved using time pulsing compared to the steady flow in all tested micromixers. In a recent study, Afzal and Kim [3] showed that a convergent-divergent channel with sinusoidal walls (space-periodic) represented the most effective coupling with pulsatile (time-periodic) flow. The mass fraction distributions show that the interfacial area increases rapidly, and separates to produce discrete puffs of fluids for enhanced mixing performance as shown in Fig. 2.1. Within two periods of the sinusoidal walls, the micromixer with pulsing inputs showed a mixing efficiency of 92%.

The micromixers based on electrokinetic flow utilize fluctuating electric field for effective mixing [4–6] as an alternative to a pressure-driven flow. The coupling of electric field and ionic conductivity results in an electric body force which generates flow instabilities, and enhances mixing. Oddy et al. [4] developed an electrokinetic process to mix solutions for bioanalytical applications. The instability caused by oscillating electroosmotic flow was used to stir fluid streams at Reynolds numbers of order unity, which was defined using the channel depth and rms electroosmotic velocity. They designed and fabricated two micromixing devices using electrokinetic instability for rapid mixing of fluid streams. Figure 2.2 shows the development of mixing with time after onset of the instability. After sufficient time, approximately homogeneous fluorescence intensity indicating good mixing is seen in the mixing chamber.

Magneto-hydrodynamics micromixers utilize a magnetic field and magnetic particles suspended in a fluid to enhance mass transport [8–11]. To improve the performance of a Y-shaped micromixer, Tsai et al. [9] and Fu et al. [10] used ferrofluid (a liquid with a suspension of magnetic nanoparticles) and permanent magnets to induce magnetoconvective flow in the channel. Hejazian et al. [11] used non-uniform magnetic field, diluted ferrofluid, and hydrodynamic flow-focusing configuration to improve mass transport in a microfluidic device shown in Fig. 2.3. The system consists of a core stream and two sheath streams. A magnetoconvective secondary flow occurred in the channel promoted the mass transport of the non-magnetic fluorescent dye.

A mixing technique based on bubble-induced acoustic microstreaming principle was developed by Liu et al. [12]. The proposed mixer included a piezoelectric disk attached to a reaction chamber, which was designed in such a way that a desired number of air bubbles with desirable size are trapped in the solution. Experiments showed that steady circulatory flows were generated through vibrations by the sound field on air bubbles resting on a solid surface, and rapid mixing was achieved (Fig. 2.4). The time for complete mixing in a 100 μL chamber was significantly reduced from hours (a pure diffusion-based mixing) to tens of seconds. Haeberle et al. [13] introduced a novel active mixing concept using centrifugal force generated by a fixed rotating drive (also known as 'player') to pump and mix the liquid educts as shown in Fig. 2.5. Fast mixing with high volume throughput was obtained as a result of induced Coriolis force.

Brief descriptions of typical designs of active micromixers and their applications can be found in reviews of Nguyen and Wu [14], Hessel et al. [15, 16] and Kumar et al. [17].

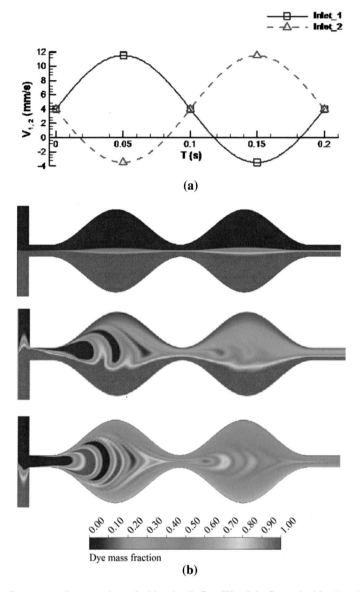

Fig. 2.1 Convergent-divergent channel with pulsatile flow [3]: **a** Inlet flow velocities ($\phi = 180°$ and $St = 0.139$), and **b** dye mass fraction distributions on the x–y plane at one half of the channel depth ($z = 0.0625$ mm) at time $t = 0/T$ in a cycle for (top) no pulsing, (middle) pulsing from one-inlet, and (bottom) pulsing from both inlets with $\varphi = 180°$ with $V_o/V_s = 1.88$ and $St = 0.278$

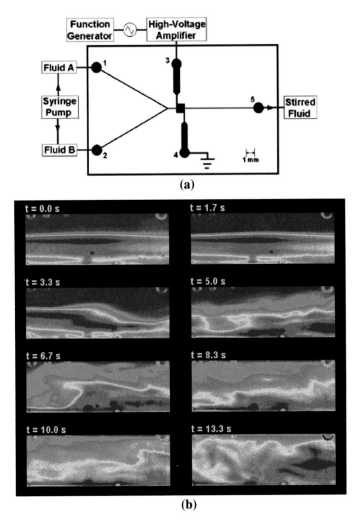

Fig. 2.2 Micromixer using oscillating electroosmotic flow [4]: **a** Schematic of micromixer based on electrokinetic instability, and **b** time-stamped images obtained from micromixer showing an initially stable interface and its development after the onset of the instability. Some air bubbles are shown clinging to the PDMS channel walls

2.2 Passive Micromixers

Passive micromixers exploit the micromixer's geometry to produce complex flow fields for effective mixing. They are free from moving parts and can be fabricated in both planar and 3D configurations.

Examples of simple passive micromixer are T- and Y- shaped microchannels shown in Fig. 2.6. T-mixers have been used by many researchers to investigate

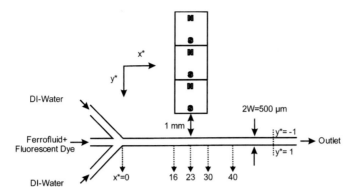

Fig. 2.3 Schematic of a magneto-hydrodynamics micromixer and a set of permanent magnets [11]

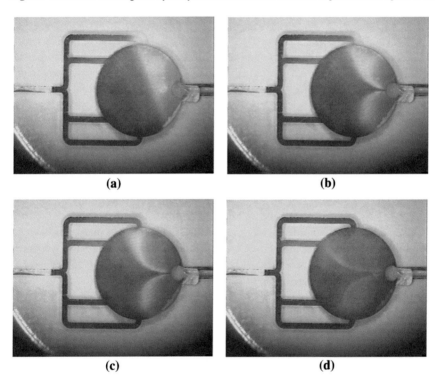

Fig. 2.4 Micromixer based on bubble-induced acoustic microstreaming—photographs (top view) showing acoustic microstreaming occurred around a single air bubble [12]: **a** time 0, **b** 15 s, **c** 35 s, and **d** 1 min and 10 s

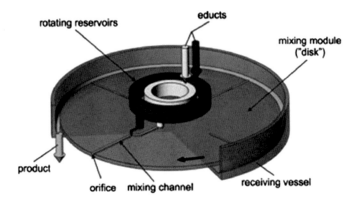

Fig. 2.5 Centrifugal micromixer [13]

fundamental mixing processes, both experimentally and numerically [18–22]. Dreher et al. [18] carried out numerical and experimental investigations to study the flow dynamics and mixing in a T-shaped micromixer over a wide Reynolds number range (0.01–1000). Engler et al. [19] and Kockmann et al. [20] found different flow regimes depending on Reynolds number in T-micromixers. In Fig. 2.6c, three distinct laminar flow regimes are observed: stratified, vortex and engulfment flows. In contrast to the stratified and vortex flows, the engulfment flow at high Reynolds numbers leads to significant improvement in mixing performance. Also, a new identification number was introduced to take into account the pressure loss in the microchannel.

Kockmann et al. [21] conducted a theoretical and experimental investigation of convective micromixing in different mixer structures, viz. asymmetric T-mixers, TTree-mixers, tangential mixers, and double T-mixers. Afzal and Kim [22] carried out numerical simulation to investigate the flow dynamics and mixing behavior of non-Newtonian working fluids in a T-shaped microchannel using shear-dependent viscosity models. The Carreau-Yasuda [23] and Casson [24] non-Newtonian blood viscosity models were used to capture the non-Newtonian flow characteristics. Under similar operating conditions, flow dynamics and mixing were evaluated for different working fluids: Newtonian fluid (water) and non-Newtonian fluid (blood) using the Carreau-Yasuda model. For low mass flow rates, the mixing performances of both the fluids were found to be nearly equivalent, and decreased with flow rate. However, for high flow rates, mixing with water significantly improved, but for blood, only a negligible change in mixing performance was observed. Ansari et al. [25] proposed a novel vortex micro T-mixer with tangentially aligned inlet channels for a wide Reynolds number range (Fig. 2.7). A vortex was formed at the inlet of the rectangular channel leading to stretching and folding of fluid interface.

A survey of the open literature shows that most passive micromixers fall under two broad categories:

1. Multi-lamination and focusing
2. Chaotic advection

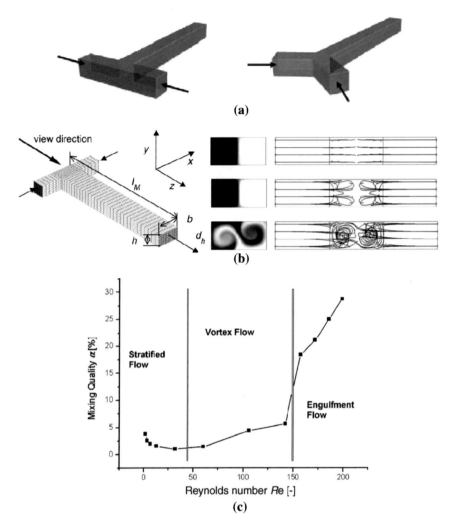

Fig. 2.6 Simple passive micromixers: **a** T- (left) and Y- (right) shaped micromixers [21], **b** numerical simulation of the fluid flow and mixing in a T-shaped micro mixer; left: geometry of the numerical model; middle: concentration distribution in the mixing channel for 1:1-mixing; right: streamlines in a T-shaped micro mixer for three flow situations: straight laminar, laminar vortex, and engulfment flow, and **c** mixing quality α over Reynolds number for a T-micromixer [19]

The idea behind multi-lamination micromixers is splitting a fluid stream into 'n' sub-streams, thereby increasing the contact surface for mixing liquids. The multi-lamination also reduces the diffusion path between the co-flowing fluid streams, and helps to improve molecular diffusion.

Hessel et al. [26] developed interdigital micromixer designs with alternating feed channels to periodically create liquid multi-lamellae for basic investigation of mixing. The micromixers were made of glass using etching techniques. The three

Fig. 2.7 Vortex T-mixer
with non-aligned inputs [25]

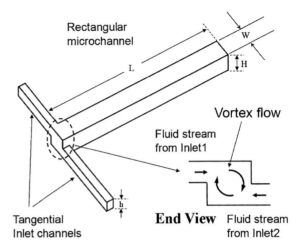

interdigital designs, i.e. rectangular, triangular, and slit-shaped designs shown in Fig. 2.8, differed in the flow-through mixing chamber. Flow patterns in both the triangular and slit-shaped micromixers differed from those in regular multi-lamination micromixers. A special version of the triangular mixer, super-focus mixer, showed that time for mixing was reduced to about 10 ms, as determined by iron-rhodanide reaction imaging. In the latter case, the lamellae were compressed by a factor of 40, from a width of 160 μm to 4 μm. The work highlighted an important aspect of focusing the inlet streams to narrow mixing channel. Veenstra et al. [27] used hydrodynamic focusing to compress fluid lamellae to reduce the diffusion distance to a few micrometers to enhance mixing.

The basic idea involved in the design of micromixers based on chaotic advection is stretching, folding and breaking of the flow. Chaotic advection can be generated using specially designed geometries of the microchannel in passive micromixers or induced by an external force in active micromixers. Types of geometrical modifications used to generate chaotic advection are:

- Surface patterning
- Serpentine channels
- Obstacles in flow channel
- Split and recombination of flow paths
- Selected combinations of above modifications

Patterned topography can be used to generate transverse flow that increases the interfacial area between the fluids to be mixed. A typical patterning method to enhance the mixing process was proposed by Stroock et al. [28] using bas-relief structures (staggered herringbone grooves) on the floor of a channel. The micromixer was developed using two-step photolithography in SU-8 photoresist. The channel structure was made using the first layer of photolithography whereas the pattern of ridges was realized in the second layer. In the staggered herringbone micromixer

Fig. 2.8 Multi-lamination micromixers: **a** Interdigital micromixer [26], and **b** super-focus micromixer [15]

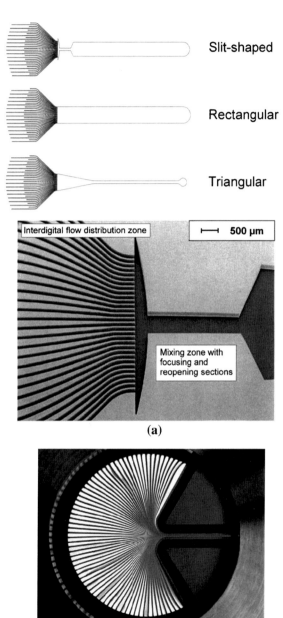

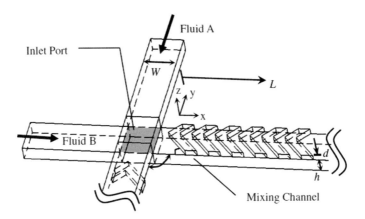

Fig. 2.9 Schematic diagram of overlapping crisscross micromixers [32]

(SHM), a series of rotational and extensional local flows were developed by varying the shape of the grooves as a function of the axial channel length, which led to better development of chaotic advection inside the micromixer and enhanced mixing performance.

Yang et al. [29] studied the effects of geometric parameters on the mixing performance of a SHM using numerical simulation and Taguchi method. The depth ratio and asymmetry index were found to be the most dominant geometric parameters affecting the overall mixing performance. Hassell and Zimmerman [30] performed a computational study of flow dynamics and mixing inside three different configurations of SHM using a finite-element based software. Wang and Yang [31] and Wang et al. [32] investigated a grooved micromixer incorporating an overlapping crisscross inlet ports located at the intersection of two patterned channels crossing one above another using numerical analysis and experiment. Figure 2.9 shows layout overlapping crisscross micromixers proposed by Wang et al. [32]. The polydimethyl-siloxane (PDMS) micromixer was fabricated using photolithographic process and cast-molding technology. The proposed design of the micromixer demonstrated superior mixing performance over the existing herringbone mixer.

In serpentine channels with rectangular or circular cross sections, secondary flows evolve as a result of centrifugal force. These are characterized by the presence of two counter-rotating vortices called Dean vortices [33, 34]. The vortices provide an effective means of enhancing mixing through the stretching and folding of fluid interfaces, thereby increasing the surface area across which diffusion occurs. Vanka et al. [35] carried out a computational study to determine mixing rates in a curved square duct at low Reynolds numbers applicable to microfluidic applications. Several cases were simulated for different Reynolds and Schmidt numbers to highlight the effect of duct curvature on secondary flow behavior. For high Schmidt number fluids, mixing could be enhanced for higher Reynolds numbers ($Re \geq 10$).

Under pressure-driven flow through a curved channel of square cross section, Howell et al. [36] showed the formation of Dean vortices. The curved channels were fabricated using machining processes in a poly-methyl methacrylate (PMMA) sheet (Lucite CP, ICI Acrylics Inc, Cordova TN) with a Techno-isel CNC router (Techno, Inc., New Hyde Park, NY). The vortices were first seen at Reynolds numbers between 1 and 10, and became stronger as the flow velocity increased. Higher aspect ratio (depth/width) channels also showed improved mixing performance. Figure 2.10 shows images of the outlet end of the channel at different Reynolds numbers. At Re

Fig. 2.10 Images of the outlet end of the curved channel at different Reynolds numbers. Channel dimensions are 1.27 mm by 1.27 mm. Radius of curvature is 5 mm. All flows are at steady state [36]

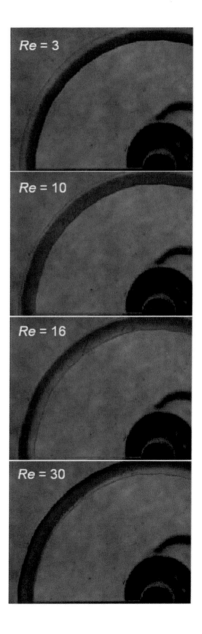

= 3, the dye, which was introduced in the inner half of the channel, moves to the outer half by the process of diffusion. On the other hand, a very rapid mixing was achieved at Re = 30 due to strong Dean vortices developed in the channel. Jiang et al. [37] used such secondary flows to induce chaotic mixing in the flows through meandering channels. The secondary motions were characterized using Dean number. For low values of Dean number, the secondary flow consisted of two counter-rotating vortices. Above a critical value of Dean number, two additional counter-rotating vortices appeared near the outer wall. The four-vortex pattern resulted in a rapid development of mixing in the microchannel. However, the operating Reynolds number should be sufficiently high (Re ~ 300) in order to utilize the full potential of the micromixer.

Liu et al. [38] studied experimentally the mixing in a 3D serpentine microchannel with C-shaped repeating units in a Reynolds number range, 6 ≤ Re ≤ 70. Using a double-sided KOH (Potassium Hydroxide) wet-etching technique, the 3D micromixer was fabricated in a silicon wafer. The mixing capability of the micromixer was found to increase with Reynolds number, and nearly complete mixing was observed at Re = 70 with the occurrence of chaotic advection. The mixing performance of the 3D serpentine channel was found to be superior to that of a planar serpentine channel.

Figure 2.11 shows planar micromixer designs, viz. square-wave, zig-zag and sinusoidal micromixers. Lin and Yang [39] conducted a computational study on a square-wave channel. Mengeaud et al. [40] investigated numerically the mixing process in a zig-zag microchannel integrated with a Y-inlet junction in 1 < Re < 800 using the finite element method. Their results characterized the effects of channel geometry and flow rate on hydrodynamics and mixing efficiency. Molecular diffusion dominated the mixing process up to a Reynolds number of about 80; however, recirculation zones developed at higher Reynolds numbers enhanced the mixing performance. Afzal and Kim [41] proposed a micromixer design with sinusoidal channel walls,

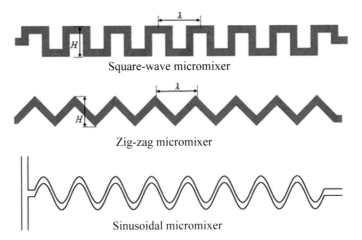

Square-wave micromixer

Zig-zag micromixer

Sinusoidal micromixer

Fig. 2.11 Planar passive micromixers [41]

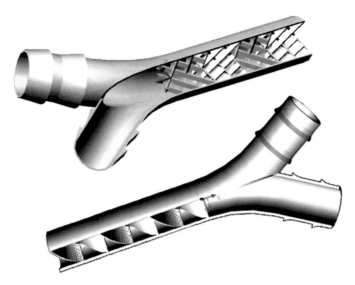

Fig. 2.12 Two different designs of micromixer based on chaotic advection [42]. (top) Micromixer made of intersecting channels. (bottom) Micromixer made of helical elements

and performed numerical simulations to study the mixing performance for a wide Reynolds number range. The proposed sinusoidal micromixer showed much better mixing performance than the square-wave and zigzag micromixers for the same wavelength.

Using micro-stereolithography, Bertsch et al. [42] realized two different designs of micromixer based on chaotic advection as shown in Fig. 2.12. The first type was composed of a series of stationary rigid elements that form intersecting channels to split, rearrange and combine component streams. The second type was composed of a series of short helix elements arranged in pairs, and each pair comprised right-handed and left-handed elements arranged alternately in a pipe. The micromixer showed a good mixing efficiency with a low pressure drop.

Bhagat et al. [43] investigated mixing in a straight microchannel with flow break-up obstructions over a wide range of flow conditions. The micromixer was fabricated by casting on an SU-8 resist mold using PDMS. The obstructions were arranged throughout the channel cross section in a repetitive fashion, which helped to break up and recombine the flow, as shown in Fig. 2.13. The mixing efficiency was found to depend on spatial arrangement and the number of obstructions. The micromixer showed good mixing performance in a low-Re range (Re < 1). Alam et al. [44] performed a numerical investigation of fluid flow and mixing performance in curved microchannels with cylindrical, hexagonal and diamond shape obstacles. The curved channel with cylindrical obstructions showed a remarkable increase in mixing performance compared to a T-channel with cylindrical obstructions and a simple curved channel in a wide Reynolds number range.

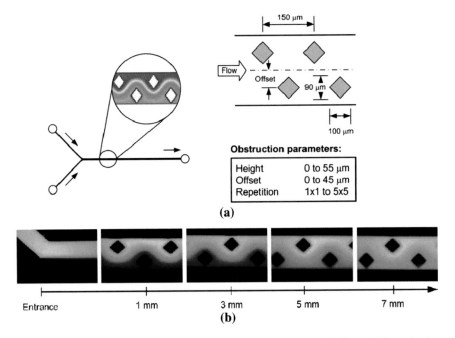

Fig. 2.13 Micromixer with flow break-up obstructions [43]: **a** Schematic diagram illustrating key features and dimensions of the micromixer, and **b** representative experimental results illustrating mixing at Re = 0.1. Fluorescein is flowing in the upper portion of the images (pseudo-colored green) and water in the lower portion (black)

Lee et al. [45, 46] developed a SAR micromixer using PDMS, and performed a numerical and experimental study to find the mixing performance in a range of mass flow rate, 0.1 μl min^{-1}–1000 μl min^{-1}. Using blue dye and water, experiment was performed to evaluate the mixing efficiency by calculating the standard deviation of the pixel intensity of the observed image (Fig. 2.14). In the SAR micromixer, interfaces were extended exponentially, and 90% mixing was obtained after 7th unit at Re = 0.6.

Ansari et al. [47, 48] studied both numerically and experimentally a planar micromixer based on unbalanced splits and cross-collisions of fluid streams shown in Fig. 2.15a in a Reynolds number range of $10 \leq$ Re ≤ 80. The micromixer was developed using PDMS replica molding method which involves preparing a SU-8 resist mold over a silicon wafer by photolithography, and transferring the pattern of the micromixer to PDMS replica. The main channel was split into two sub-channels of unequal widths to create unbalanced collisions. The mixing was mainly due to the combined effects of unbalanced collisions and Dean vortices. The unbalanced collisions proved to be more effective than the balanced collisions of fluid streams in both rhombic and circular channels. In a recent study, Afzal and Kim [49] proposed SAR micromixers with convergent-divergent sinusoidal walls (M1 and M2) as shown in

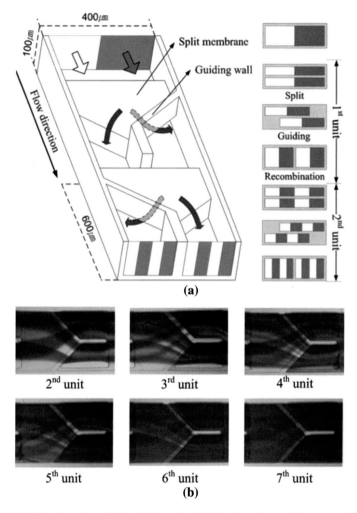

Fig. 2.14 A SAR micromixer [45]: **a** Schematic of SAR micromixer and corresponding cross-sectional view of the flow, and **b** results of the mixing experiment involving blue dye and water

Fig. 2.15b. The proposed micromixers showed significantly improved mixing performance compared to a micromixer of the similar geometry based on the concept of unbalanced splits and collisions in a wide range of Reynolds numbers. Also, the pressure drop was acceptable for lab-on-a-chip and μ-TAS.

Sudarsan and Ugaz [50, 51] developed two different micromixers shown in Fig. 2.16: (a) planar SAR (P-SAR) micromixer capable of generating alternating lamellae of individual fluid species in a SAR arrangement, and (b) multi-vortex micromixer based on coupled Dean and expansion vortices (arising out due to abrupt change in cross-sectional area). Tafti et al. [52, 53] proposed an in-plane

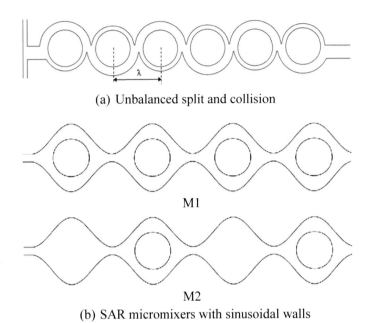

(a) Unbalanced split and collision

M1

M2

(b) SAR micromixers with sinusoidal walls

Fig. 2.15 Micromixers based on SAR mechanism [49]

passive micromixing concept using a sigma micromixer shown in Fig. 2.17. Diffusive mass flux between two miscible streams, flowing laminar in the microchannel, was enhanced when the velocity at the diffusion interface was increased. Microparticle image velocimetry was utilized for visualization of the flow. The result showed the effectiveness of the proposed passive micromixing concept and improvement of mixing performance compared to existing designs.

Using conventional UV-lithography, Hong et al. [54] designed, fabricated and successfully characterized a passive micromixer with modified Tesla structure shown in Fig. 2.18. At low flow rates, the mixing is primarily by diffusion, but at high flow rates both convection and diffusion contributes to mixing. Excellent mixing performance over a wide range of flow conditions at micro scale was observed using simulation and experiment. Chung et al. [55] proposed a passive micromixer design that utilizes the self-circulation of the fluids in the mixing chamber as shown in Fig. 2.19. The micromixer consisted of an inlet port, a circular mixing chamber and an outlet port. The micromixer was constructed with two-layers PMMA. The upper PMMA layer was blank. Using a CNC high-speed engraving and milling machine, the micromixer structures were built on the lower PMMA layer. The self-circulation phenomenon in the micromixer was studied using numerical simulation. At Re = 10, no circulation zone was observed, and the flow was similar to the symmetrical creeping flow. However, at Re = 300, a large self-circulation area was seen in the mixing chamber, which promotes mixing.

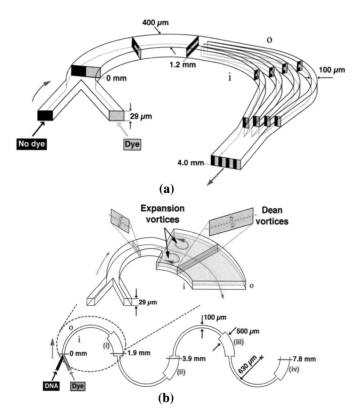

Fig. 2.16 Two planer micromixers proposed by Sudarsan and Ugaz [51]: **a** P-SAR micromixer, and **b** multi-vortex micromixer (Copyright (2006), National Academy of Sciences, U.S.A.)

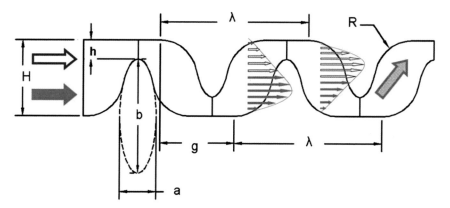

Fig. 2.17 Sigma micromixer [52]

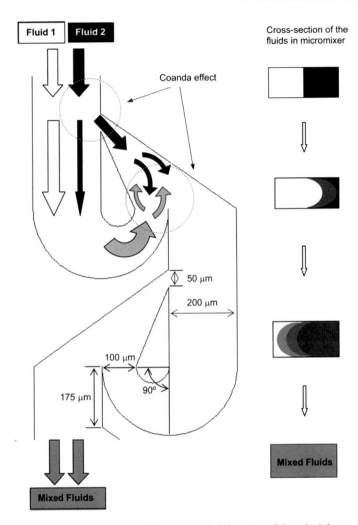

Fig. 2.18 Modified Tesla micromixer [54]—schematic illustration of the principle

Owing to the development of microfabrication techniques, 3D micromixer designs become popular. 3D micromixers have complicated the geometries, but showed generally high mixing performance. Nonetheless, the trade-off between the difficulty in 3D fabrication and high mixing performance need to be considered. Xia et al. [56] proposed 3D passive micromixer designs operating at low Reynolds numbers. The micromixers, known as two-layer crossing channel micromixers (TLCCMs), were realized as TLCCM-model A and TLCCM-model B shown in Fig. 2.20. Using laser ablation method, the micromixers were made of 1.5 mm thick transparent polymer poly (methyl methacrylate) plates. They found that chaotic advection

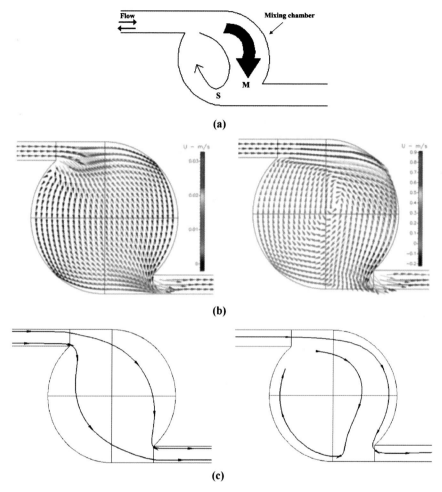

Fig. 2.19 Micromixer with circular mixing chambers [55]: **a** Schematic diagram of the two regions in the chamber, main flow (M) and circulation, **b** calculated velocity field, and **c** streamlines in the mixer with the circular chamber at Re equals (left) 10 and (right) 300

(stretching, folding and splitting, reorientation and recombination) resulted in significant improvement in mixing performance for both models A and B compared to a simple 3D serpentine microchannel. In the proposed micromixers, chaotic advection was observed even at low Reynolds numbers much less than 1.

A 3D SAR micromixer composed of O- and H- shaped units shown in Fig. 2.21 was proposed, and numerically investigated by Hossain and Kim [57]. The split and recombination of fluid streams due to repeated O- and H-shaped units along the channel length promoted chaotic advection, and improved mixing performance. For Re ≥ 30, the micromixer with ten O- and H-units showed complete mixing at the exit of the channel. In a later study, Raza et al. [58] proposed a 3D serpentine crisscross

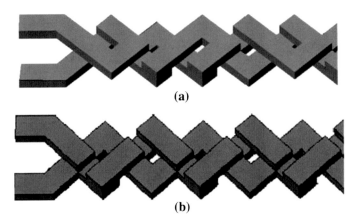

(a)

(b)

Fig. 2.20 Configurations of two-layer crossing channel micromixer (TLCCM). **a** Model A and **b** Model B [56]

Fig. 2.21. 3D serpentine SAR micromixer [57]

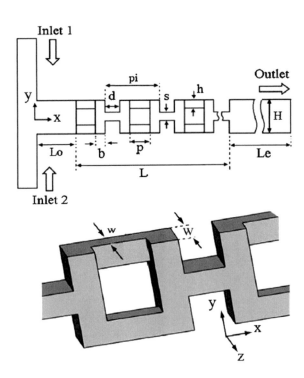

SAR design with repeating two-layer O- and X-shaped units shown in Fig. 2.22. The micromixer achieved complete mixing over a short mixing length of 1.5 mm. The high mixing performance of the micromixer was attributed to saddle-shaped flow structure which promoted chaotic advection. These 3D micromixers [56–58] showed high mixing performances over a wide Reynolds number range, and significantly

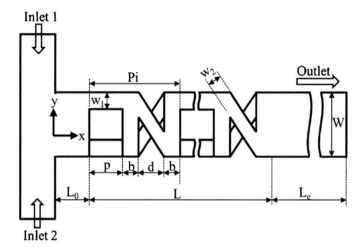

Fig. 2.22. 3D serpentine crisscross SAR micromixer with repeating OX-shaped units [58]

better mixing performance compared to planar micromixers. The performances of 3D micromixers were reviewed by Raza et al. [59] with a comparative analysis.

In summary, the selection of a particular micromixer depends on many factors: mixing index, pressure drop, simple/complicated design, operational Reynolds number, fabrication techniques and application. Planar passive micromixers [36, 37, 39–41, 47, 49, 54, 55] have simple designs, which are easy to be fabricated using conventional techniques, and offer an advantage of integration to lab-on-a-chip systems. Most planar designs rely on secondary flows for effective mixing, and therefore the pressure drop becomes an important criterion for model selection. On the other hand, 3D micromixers can be expensive due to their complicated design and fabrication complexity, but provide good mixing [38, 45, 46, 56–58].

References

1. Glasgow I, Aubry N (2003) Enhancement of microfluidic mixing using time pulsing. Lab Chip 3:114–120. https://doi.org/10.1039/B302569A
2. Glasgow I, Lieber S, Aubry N (2004) Parameters influencing pulsed flow mixing in microchannels. Anal Chem 76:4825–4832. https://doi.org/10.1021/ac049813m
3. Afzal A, Kim KY (2015) Convergent-divergent micromixer coupled with pulsatile flow. Sens Actuators B Chem 211:198–205. https://doi.org/10.1016/j.snb.2015.01.062
4. Oddy MH, Santiago JG, Mikkelsen JC (2001) Electrokinetic instability micromixing. Anal Chem 73:5822–5832. https://doi.org/10.1021/ac0155411
5. Jacobson SC, McKnight TE, Ramsey JM (1999) Microfluidic devices for electro kinematically driven parallel and serial mixing. Analyt Chem 71:4455–4459. https://doi.org/10.1021/ac990576a
6. Moctar AOE, Aubry N, Batton J (2003) Electro-hydrodynamic micro-fluidic mixer. Lab Chip 3:273–280. https://doi.org/10.1039/B306868B

7. Huang MZ, Yang RJ, Tai CH, Tsai CH, Fu LM (2006) Application of electro kinetic instability flow for enhanced micromixing in cross-shaped microchannel. Biomed Microdevice 8:309–315. https://doi.org/10.1007/s10544-006-0034-z
8. Bau HH, Zhong J, Yi M (2001) A minute magneto hydro dynamic (MHD) mixer. Sens Actuat B: Chem 79:207–215. https://doi.org/10.1016/S0925-4005(01)00851-6
9. Tsai TH, Liou DS, Kuo LS, Chen PH (2009) Rapid mixing between ferro-nanofluid and water in a semi-active Y-type micromixer. Sens Actuators A 153:267–273. https://doi.org/10.1016/j.sna.2009.05.004
10. Fu LM, Tsai CH, Leong KP, Wen CY (2010) Rapid micromixer via Ferro fluids. 12th international conference on magnetic fluids. Phys Procedia 9:270–273
11. Hejazian M, Phan DT, Nguyen NT (2016) Mass transport improvement in microscale using diluted ferrofluid and a non-uniform magnetic field. RSC Adv 6:62439–62444. https://doi.org/10.1039/C6RA11703A
12. Liu RH, Yang J, Pindera MZ, Athavale M, Grodzinski P (2002) Bubble induced acoustic micromixing. Lab Chip 2:151–157. https://doi.org/10.1039/B201952C
13. Haeberle S, Brenner T, Schlosser HP, Zengerle R, Ducrée J (2005) Centrifugal micromixer. Chemical. Eng Technol 28:613–616. https://doi.org/10.1002/ceat.200407138
14. Nguyen NT, Wu Z (2005) Micromixers-a review. J Micromech Microeng 15:R1–R16. https://doi.org/10.1088/0960-1317/15/2/R01
15. Hardt S, Drese KS, Hessel V, Schönfeld F (2005) Passive micromixer for applications in the microreactor and μTAS fields. Microfluid Nanofluid 1:108–118. https://doi.org/10.1007/s10404-004-0029-0
16. Hessel V, Lowe H, Schönfeld F (2005) Micromixers—a review on passive and active mixing principles. Chem Eng Sci 60:2479–2501. https://doi.org/10.1016/j.ces.2004.11.033
17. Kumar V, Paraschivoiu M, Nigam KDP (2011) Single-phase fluid flow and mixing in microchannels. Chem Eng Sci 66:1329–1373. https://doi.org/10.1016/j.ces.2010.08.016
18. Dreher S, Kockmann N, Woias P (2009) Characterization of laminar transient flow regimes and mixing in T-shaped micromixers. Heat Transfer Eng 30(1–2):91–100. https://doi.org/10.1080/01457630802293480
19. Engler M, Kockmann N, Kiefer T, Woias P (2004) Numerical and experimental investigations on liquid mixing in static micromixers. Chem Eng J 101:315–322. https://doi.org/10.1016/j.cej.2003.10.017
20. Kockmann N, Engler M, Haller D, Woias, (2005) Fluid dynamics and transfer processes in bended microchannels. Heat Transfer Eng 26:71–78. https://doi.org/10.1080/01457630590907310
21. Kockmann N, Kiefer T, Engler M, Woias P (2006) Convective mixing and chemical reactions in microchannels with high flow rates. Sens. And Act B 117:495–508. https://doi.org/10.1016/j.snb.2006.01.004
22. Afzal A, Kim KY (2014) Flow and Mixing analysis of non-Newtonian fluids in straight and serpentine microchannels. Chem Eng Sci 116:263–274. https://doi.org/10.1016/j.ces.2014.05.021
23. Abraham F, Behr M, Heinkenschloss M (2005) Shape optimization in steady blood flow: a numerical study of non-Newtonian effects. Comput Methods Biomech Biomed Eng 8(2):127–137. https://doi.org/10.1080/10255840500180799
24. Perktold K, Resch, M, Florian, M (1991) Pulsatile non-Newtonian flow characteristics in a three-dimensional human carotid bifurcation model. J Biomech Eng 113(4):464–475. https://doi.org/10.1115/1.2895428
25. Ansari MA, Kim KY, Anwar K, Kim SM (2012) Vortex micro T-mixer with non-aligned inputs. Chem Eng J 162:760–767. https://doi.org/10.1016/j.cej.2011.11.113
26. Hessel V, Hardt S, Löwe H, Schönfeld F (2003) Laminar mixing in different interdigital micromixers, I: experimental characterization. AIChE J 49(3):566–577. https://doi.org/10.1002/aic.690490304
27. Veenstra TT, Lammerink TS, Elwenspoek MC, van den Berg A (1999) Characterization method for a new diffusion mixer applicable in micro flow injection analysis systems. J Micromech Microeng 9:199–202. https://doi.org/10.1088/0960-1317/9/2/323

28. Stroock AD, Dertinger SK, Ajdari A, Mezić I, Stone HA, Whitesides GM (2002) Chaotic mixer for microchannels. Science 295:647–651. https://doi.org/10.1126/science.1066238
29. Yang JT, Huang KJ, Lin YC (2005) Geometric effects on fluid mixing in passive grooved micromixers. Lab Chip 5:1140–1147. https://doi.org/10.1039/B500972C
30. Hassell DG, Zimmerman WB (2006) Investigation of the convective motion through a staggered herringbone micromixer at low Reynolds number flow. Chem Eng Sci 61:2977–2985. https://doi.org/10.1016/j.ces.2005.10.068
31. Wang L, Yang JT (2006) An overlapping crisscross micromixer using chaotic mixing principles. J Micromech Microeng 16:2684. https://doi.org/10.1088/0960-1317/16/12/022
32. Wang L, Yang JT, Lyu PC (2007) An overlapping crisscross micromixer. Chem Eng Sci 62:711–720. https://doi.org/10.1016/j.ces.2006.09.048
33. Dean WR (1927) Note on the motion of fluid in a curved pipe. Philos Mag 4:208–223. https://doi.org/10.1080/14786440708564324
34. Dean WR (1928) The stream-line motion of fluid in a curved pipe. Philos Mag 5:673–695. https://doi.org/10.1080/14786440408564513
35. Vanka SP, Luo G, Winkler CM (2004) Numerical study of scalar mixing in curved channels at low Reynolds number. AIChE J 50:2359–2368. https://doi.org/10.1002/aic.10196
36. Howell PB Jr, Mott DR, Golden JP, Ligler FS (2004) Design and Evaluation of a Dean Vortex-Based Micromixer. Lab Chip 4:663–669. https://doi.org/10.1039/B407170K
37. Jiang F, Drese KS, Hardt S, Küpper M, Schönfeld F (2004) Helical flows and chaotic mixing in curved micro channels. AIChE J. 50:2297–2305. https://doi.org/10.1002/aic.10188
38. Liu RH, Stremler MA, Sharp KV, Olsen MG, Santiago JG, Adrian RJ, Aref H, Beebe DJ (2000) Passive mixing in a three-Dimensional serpentine microchannel. J Microelectromech Syst 9:190–197. https://doi.org/10.1109/84.846699
39. Lin KW, Yang JT (2007) Chaotic mixing of fluids in a planar serpentine channel. Int J Heat Mass Transf 50:1269–1277. https://doi.org/10.1016/j.ijheatmasstransfer.2006.09.016
40. Mengeaud V, Josserand J, Girault HH (2002) Mixing processes in a zigzag microchannel: finite element simulations and optical study. Anal Chem 74:4279–4286. https://doi.org/10.1021/ac025642e
41. Afzal A, Kim KY (2013) Mixing Performance of a passive micromixer with sinusoidal channel walls. J Chem Eng Jpn 46:230–238. https://doi.org/10.1252/jcej.12we144
42. Bertsch A, Heimgartner S, Cousseau P, Renaud P (2001) Static mixers based on large-scale industrial mixer geometry. Lab Chip 1:56–60. https://doi.org/10.1039/B103848F
43. Bhagat AA, Peterson ET, Papautsky I (2010) A passive planar micromixer with obstructions for mixing at low Reynolds number. J Micromech Microeng 20:1–10. https://doi.org/10.1088/0960-1317/17/5/023
44. Alam A, Afzal A, Kim KY (2014) Mixing Performance of a planar micromixer with circular obstructions in a curved microchannel. Chem Eng Res Des 92:423–434. https://doi.org/10.1016/j.cherd.2013.09.008
45. Lee SW, Kim DS, Lee SS, Kwon TH (2006) A split and recombination micromixer fabricated in a PDMS three-dimensional structure. J Micromach Microeng 16:1067–1072. https://doi.org/10.1088/0960-1317/16/5/027
46. Lee SW, Lee SS (2008) Rotation effect in split and recombination micromixing. Sens Actuat B 129:364–371. https://doi.org/10.1016/j.snb.2007.08.038
47. Ansari MA, Kim KY, Anwar K, Kim SM (2010) A novel passive micromixer based on unbalanced splits and collisions of fluid streams. J Micromech Microeng 20:1–10. https://doi.org/10.1088/0960-1317/20/5/055007
48. Ansari MA, Kim KY (2010) Mixing performance of unbalanced split and recombine micromixers with circular and rhombic sub-channels. Chem Eng J 162:760–767. https://doi.org/10.1016/j.cej.2010.05.068
49. Afzal A, Kim KY (2012) Passive split and recombination micromixer with convergent-divergent walls. Chem Eng J 203:182–192. https://doi.org/10.1016/j.cej.2012.06.111
50. Sudarsan AP, Ugaz VM (2006a) Fluid mixing in planar spiral microchannels. Lab Chip 6:74–82

51. Sudarsan AP, Ugaz VM (2006b) Multivortex micromixing. Proc Natl Acad Sci 103(19):7228–7233
52. Yakhshi-Tafti E, Kumar R, Cho HJ (2008) Effect of laminar velocity profile variation on mixing in microfluidic devices: the sigma micromixer. Appl Phys Lett 93:1–3. https://doi.org/10.1063/1.2996564
53. Yakhshi-Tafti E, Cho HJ, Kumar R (2011) Diffusive mixing through velocity profile variation in microchannels. Exp Fluids 50:535–545. https://doi.org/10.1007/s00348-010-0954-5
54. Hong CC, Choi JW, Ahn CH (2004) A novel in-plane passive microfluidic mixer with modified Tesla structures. Lab Chip 4:109–113. https://doi.org/10.1039/B305892A
55. Chung YC, Hsu YL, Jen CP, Lu MC, Lin YC (2004) Design of passive mixers utilizing microfluidic self-circulation in the mixing chamber. Lab Chip 4:70–77. https://doi.org/10.1039/B310848C
56. Xia HM, Wan SYM, Shu C, Chew YT (2005) Chaotic micromixers using two-layer crossing channels to exhibit fast mixing at low Reynolds numbers. Lab Chip 5:748–755. https://doi.org/10.1039/b502031j
57. Hossain S, Kim KY (2015) Mixing Analysis in a Three-Dimensional Serpentine Splitl Engineering Research and Design 100:95–103. https://doi.org/10.1016/j.cherd.2015.05.011
58. Raza W, Hossain S, Kim KY (2018) Effective mixing in a short serpentine split-and-recombination micromixer. Sens Actuat B Chem 258:381–392. https://doi.org/10.1016/j.snb.2017.11.135
59. Raza W, Hossain S, Kim KY (2020) A review of passive micromixers with a comparative analysis. Micromachines 11(5):455–480. https://doi.org/10.3390/mi11050455

Chapter 3
Computational Analysis of Flow and Mixing in Micromixers

Abstract This chapter introduces the computational framework and provides a detailed analysis on the different numerical techniques for the analyses of flow and mixing in micromixers. Flow and mixing analyses are based on both the Eulerian and Lagrangian approaches; relative advantages and disadvantages of these two approaches and suitability to different types of mixing problems are analyzed. This chapter also discusses the various facets of numerical schemes subjected to discretization errors and computational grid requirements. Since a large number of studies are based on commercial CFD packages, relevant details of these packages to the mixing problem are presented. This chapter concludes with mixing characterization technique using concentration data obtained on a computational grid, and provides the basis for performance evaluation of different micromixer designs. This chapter consists of three sections. Section 3.1 presents the Eulerian approach for flow and mixing analyses, different mixing models, boundary conditions, and the numerical approach employed in obtaining solutions of the governing equations. The Lagrangian approach is presented in Sect. 3.2. In the final section, the method for mixing quantification is discussed.

3.1 Eulerian Approach to Mixing

The following assumptions are made to obtain the simplified governing equations for flow dynamics in micromixers:

1. Smooth and isothermal walls
2. Incompressible and Newtonian fluid flow
3. Constant diffusion coefficient for the two mixing fluids
4. Negligible wall surface tension

The continuity and Navier–Stokes equations can be written as follows:

$$\frac{\partial U_i}{\partial x_i} = 0 \tag{3.1}$$

© The Author(s), under exclusive license to Springer Nature Singapore Pte Ltd. 2021
A. Afzal and K.-Y. Kim, *Analysis and Design Optimization of Micromixers*,
SpringerBriefs in Computational Mechanics,
https://doi.org/10.1007/978-981-33-4291-0_3

$$\rho\left(\frac{\partial U_i}{\partial t} + U_j \frac{\partial U_i}{\partial x_j}\right) = -\frac{\partial p}{\partial x_i} + \mu \frac{\partial}{\partial x_j}\left[\left(\frac{\partial U_i}{\partial x_j} + \frac{\partial U_j}{\partial x_i}\right)\right] \qquad (3.2)$$

where U_i represents the fluid velocity, ρ is the fluid density, and μ is the dynamic viscosity. The mixing analysis is carried out with the following advection–diffusion model for the species concentration field:

$$\frac{\partial C}{\partial t} + U_j \frac{\partial C}{\partial x_j} = D \frac{\partial^2 C}{\partial^2 x_j} \qquad (3.3)$$

where D is the diffusivity coefficient and C is the concentration of the species.

The most widely used approach to analyze the performance of micromixers is to label one of the fluids with a dye (imagine adding a small amount of dye to one of the fluids), and in this case, C represents the dye mass fraction. In this model of mixing analysis, the fluids with and without dye have the same viscosity, μ, and the same fluid density, ρ. It is assumed that the variation in mass fraction of dye do not modify the viscosity and density of the fluid. Physically, Eq. (3.3) describes the behavior of a solute being passively convected by the local fluid velocity and diffused by molecular diffusion. The boundary conditions are selected carefully to match the experimental setup. A fluid (for example, pure water) enters at an inlet (mass fraction equals 0), and the other fluid (for example, solution of dye in water) enters at the other inlet (mass fraction equal to 1); both the fluids enter at constant velocities. Zero static pressure is specified at the outlet. A no-slip condition is applied at the walls.

Modeling of 3D laminar mixing under the assumption of constant fluid density and viscosity has been described in many publications, and validated experimentally for different micromixers [1–10]. Using Rhodamine B in water (diffusion coefficient, $D = 2.8 \times 10^{-10}$ m^2 s^{-1}, Sc = 3588), Kockmann et al. [4] studied convective mixing in microchannels. Mengeaud et al. [5] performed numerical simulations of flow and mixing in a zig-zag micromixer using properties of water and diffusion coefficient ranging from 10^{-9} to 10^{-6} m^2 s^{-1}. Tsai and Wu [8] numerically investigated mixing in a curved-straight-curved (CSC) micromixer using CFD-ACE + software. The working fluid was deionized (DI) water, and the diffusion coefficient of fluorescent dye in water was set to 3.6×10^{-10} m^2 s^{-1}. Figure 3.1 compares qualitatively their numerical results with experimental images [8] at different Reynolds numbers, Re = 1, 9 and 81. It can be seen that the numerically predicted concentration distributions are in good agreement with the experimental images. To study flow and mixing in a sigma micromixer, Afzal and Kim [10] used water and dye water at 25 °C. As can be seen from Fig. 3.2, their numerical results are in good agreement with the experimental data [11], and successfully captures the trend in mixing performance.

Another approach is to use multi-component model which estimates both density and viscosity of the local flow from mass fractions of the mixing fluids. This approach considers the mixing of two fluids with different densities and viscosities such as a water–ethanol system. The values of density and viscosity for water

Fig. 3.1 Concentration distribution on the horizontal mid-plane of the micromixer: (left) numerical simulations, (right) experimental photographs at **a** Re = 1, **b** Re = 9, and **c** Re = 81 [8]

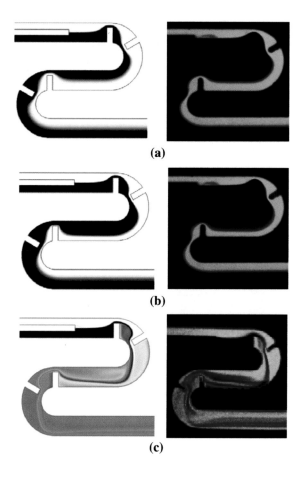

(a)

(b)

(c)

are 998 kg m^{-3} and 10^{-3} kg m^{-1} s^{-1} respectively. The corresponding values for ethanol are 789 kg m^{-3} and 1.2 × 10^{-3} kg m^{-1} s^{-1} respectively. The diffusivity coefficient for water–ethanol pair is 1.2 × 10^{-9} m^2 s^{-1}. The properties of ethanol and water were measured at 20 °C. A multi-component model available in ANSYS - CFX® [12] has been used by many authors for analyses of flow and mixing in micromixers [13–16]. The solver calculates the appropriate average values of the properties (density and viscosity) at each local control volume in the flow domain to calculate the flow. Thus, these average values depend on both the component property values and the proportion of each component at the location. Details on the theory and implementation of the multi-component model can be found in ANSYS - CFX® solver theory guide [12].

In this type of modeling, the most important thing is to determine the dependency of viscosity, μ, and density, ρ on species concentration. Orsi et al. [16] conducted a numerical investigation of flow and mixing in a T-shaped micromixer for a water–ethanol system using a commercial CFD package, ANSYS–Fluent 12.0®. Instead of

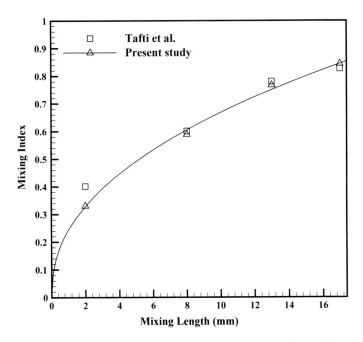

Fig. 3.2 Variations in mixing index with mixing length for sigma micromixer at Re = 0.91 [10]

using the linear functional dependences of density and viscosity on species concentration (default setting in ANSYS–Fluent®), functional approximations for both density and viscosity of the system were obtained by fitting experimental measurements [17]. It was observed that the linear dependence worked well for density, but was completely wrong for viscosity of the water–ethanol system.

Wu et al. [18] studied mixing in a planar passive micromixer using glycerol-water mixture. The density of the mixture was determined using a linear relationship based on the mass fractions of glycerol and water, but viscosity and diffusion coefficient of the mixture were estimated using non-linear functional relationships based on the mass fraction of glycerol. A comparison with experimental data showed that the nonlinear approach achieved higher accuracy, and thus the nonlinear approach is more suitable to simulate the viscous mixing than the linear approximation. In summary, it becomes important to introduce the real dependence of fluid properties (viz. density, viscosity and corresponding diffusion constant) on the concentration into the numerical model in order to achieve high accuracy in the species concentration prediction.

For flow and mixing analyses in micromixers, both tetrahedral and hexahedral grids have been used in many studies [1–10, 13–16]. Some examples of grid system employed for different micromixers are shown in Fig. 3.3. Irrespective of the choice of grid system, a grid-dependency test is mandatory to fix the grid cell sizes and distribution for the accurate prediction. Even though successful grid convergence

Fig. 3.3 Examples of two
different grids: (Top)
hexahedral grid [2] and
(bottom) tetrahedral grid [38]

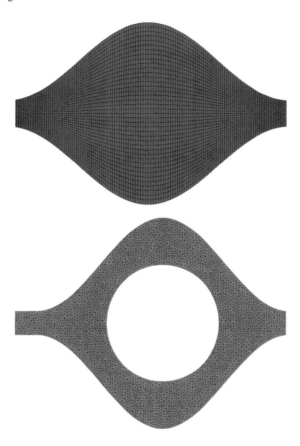

is obtained for the velocity field, discrepancies may still exist for the concentration field due to numerical (or false) diffusion, particularly at large Peclet numbers.

False diffusion can occur in the numerical solution with a poor quality grid, resulting in overprediction of mixing. Effect of false diffusion on numerical solution was tested by Liu [19] for both tetrahedral and hexahedral grids in a micromixer designed by Chen and Meiners [20]. It was found that false diffusion in the numerical solution was more pronounced for tetrahedral cells compared to hexahedral cells, and the difference was almost one order of magnitude. He recommended to avoid tetrahedral cells for mixing analysis. Also, hexahedral grids need much smaller numbers of cells, and significantly reduce the computational burden to obtain solutions compared to tetrahedral grids.

The governing equations are discretized to obtain the system of algebraic equations in a finite-element or finite-volume framework. Choice of discretization scheme affects the quality of the numerical solution. More details of the discretization and solution procedure can be found in Patankar [21] and Ferziger and Peric [22]. On the accuracy of numerical schemes for scalar mixing, Liu [19] showed that higher-order discretization schemes are less susceptible to numerical diffusion. Standard upwind differencing scheme is likely to introduce considerable discretization error in the analysis of complex flows. However, higher-order upwind (second- and third-order accurate) schemes tend to reduce numerical diffusion. Kim and coworkers [2, 10, 14, 23, 24] conducted flow and mixing analyses in different micromixers. They used a high resolution scheme to discretize the advection terms in the governing equations. The scheme reduced the numerical discretization errors using an automatic correction algorithm [12].

The linearized algebraic system of equations resulting from discretization can be solved using direct and iterative methods. However, the use of a direct method is usually not economical, and the iterative methods are more widely used. Multigrid accelerated incomplete lower–upper (ILU) factorization procedure was used in many studies. Hong et al. [1] used conjugate gradient with preconditioning solver, and convergence limit for mass fraction set to 10^{-6} for mixing study in a mixer with modified Tesla structure. Glatzel et al. [25] performed a comparative evaluation of commercial CFD packages for microfluidic applications. Four commercial finite volume codes, CFD-ACE + ®, ANSYS-CFX®, ANSYS-Fluent® and Flow-3D® have been tested to study the flow and mixing in a SAR micromixer and a rotating microchannel. For the numerical solution of the advection–diffusion equation, all the tested codes showed good predictions qualitatively, but suffered from numerical diffusion. However, any quantitative assessment of them by comparing with experimental data was not performed.

3.2   Lagrangian Approach to Mixing

As explained earlier, the numerical solution of the advection–diffusion equation for concentration field can be affected by discretization errors through numerical diffusion. To avoid this problem, Lagrangian particle tracking has been used by many researchers to study mixing in different micromixers [26–32].

In the Lagrangian formulation, a large number of particles are introduced at the inlet of the micromixer, and tracked through the flow field for the continuous phase obtained from the solution of governing Eqs. (3.1) and (3.2). The effect of particles on the flow field is negligible (one-way coupling). Movement of a massless particle inside the flow is determined by integrating the vector equation of motion for the particle:

$$\frac{d\boldsymbol{x}_p}{dt} = \mathbf{V}_p \tag{3.4}$$

For the massless particles, the particle velocity is equal to the velocity of the continuous phase. Hence, the trajectory of each particle can be obtained using the particle velocity $\mathbf{V_p} = \mathbf{V}$, where $\mathbf{V_p}$ and $\mathbf{V}$ are velocities of the particle and continuous phase, respectively. The new particle location along the trajectory is calculated using Eq. (3.4).

   Bertsch et al. [27] used particle tracking to get quantitative indication of mixing together with experiments. Wang et al. [28] numerically investigated mixing in a microchannel with patterned grooves using particle tracking algorithm based on fourth-order adaptive Runge–Kutta integration scheme. Poincaré maps were developed and analyzed to study the phenomenon of chaotic advection. Aubin et al. [29, 30] performed numerical simulations to compute the flow field for continuous phase in a micromixer, and 2,480 evenly distributed particles placed on right-hand side of the mixer inlet were tracked using 4th-order Runge–Kutta scheme with adaptive step size. Using the particle tracking data, mixing analysis was presented with spatial distribution of particles on a cross-sectional plane, maximum striation thickness and residence time distributions.

   Jiang et al. [26] constructed Poincaré maps from the data collected by tracking a large number of particles in a curved microchannel. The velocity field was obtained from numerical simulation, and the particles were tracked using second-order Runge–Kutta scheme with adaptive step-size control with a commercial post-processing software, Fieldview9® (Intelligent Light). A Lagrangian particle tracking method was used to calculate the trajectories of the massless fluid particles inside the flow using ANSYS-Fluent 12.1 [33] by Afzal and Kim [32] for design optimization of a staggered herringbone micromixer. In order to integrate the equation of motion, a combination of implicit Euler and 5th-order Runge–Kutta (derived from Cash and Karp [33]) schemes with adaptive step sizes was employed using an embedded error control. The algorithm switches between the lower (implicit Euler) and higher (5th-order Runge–Kutta) order schemes for better solution accuracy and stability. This method was employed to study the advection of particles inside the flow.

3.3 Mixing Quantification

A variance-based method has been widely employed to evaluate the mixing performance of micromixers [2, 4, 6, 23, 24, 34–38]. The variance of the species is determined on a cross-sectional plane perpendicular to the stream-wise direction. Variance is based on the concept of the intensity of segregation, which in turn is based on the variance of concentration in relation to the mean concentration. The variance of the mass fraction of the mixture on a cross-sectional plane normal to the flow direction can be expressed mathematically as:

$$\sigma = \sqrt{\frac{1}{n}\sum_{i=1}^{n}(C_i - C_m)^2} \tag{3.5}$$

where n is the number of points on the plane, C_i is the mass fraction at point i, and C_m is the optimal mixing mass fraction (= 0.5, the mass fraction in the targeted case of equal mixing of the two fluids). If a sample plane is used, the values at the sampling points are calculated by interpolating values at adjacent computational nodes. For better accuracy, the number of sampling points must be relatively high with respect to the number of nodes. It is to be noted that $0 < \sigma < 0.5$.

The mixing index is defined as:

$$M = 1 - \frac{\sigma}{\sigma_{max}} \qquad (3.6)$$

The variance is taken to be the maximum for completely unmixed fluids ($\sigma_{max} = 0.5$) and the minimum for completely mixed fluids. The mixing index varies from 0 (0% mixing) to 1 (100% mixing). A higher mixing index indicates a more homogeneous concentration and better mixing performance.

References

1. Chung YC, Hsu YL, Jen CP, Lu MC, Lin YC (2004) Design of passive mixers utilizing microfluidic self-circulation in the mixing chamber. Lab Chip 4:70–77. https://doi.org/10.1039/B310848C
2. Afzal A, Kim KY (2015) Convergent-divergent micromixer coupled with pulsatile flow. Sens Actuators B Chem 211:198–205. https://doi.org/10.1016/j.snb.2015.01.062
3. Glasgow I, Aubry N (2003) Enhancement of microfluidic mixing using time pulsing. Lab Chip 3:114–120. https://doi.org/10.1039/B302569A
4. Kockmann N, Kiefer T, Engler M, Woias P (2006) Convective mixing and chemical reactions in microchannels with high flow rates. Sens Act B 117:495–508. https://doi.org/10.1016/j.snb.2006.01.004
5. Mengeaud V, Josserand J, Girault HH (2002) Mixing processes in a zigzag microchannel: finite element simulations and optical study. Anal Chem 74:4279–4286. https://doi.org/10.1021/ac025642e
6. Bhagat AA, Peterson ET, Papautsky I (2010) A passive planar micromixer with obstructions for mixing at low Reynolds number. J Micromech Microeng 20:1–10. https://doi.org/10.1088/0960-1317/17/5/023
7. Hong CC, Choi JW, Ahn CH (2004) A novel in-plane passive microfluidic mixer with modified Tesla structures. Lab Chip 4:109–113. https://doi.org/10.1039/B305892A
8. Tsai RT, Wu CY (2011) An efficient micromixer based on multidirectional vortices due to baffles and channel curvature. Biomicrofluidics 5:1–13. https://doi.org/10.1063/1.3552992
9. Chung CK, Shih TR (2008) Effect of geometry on fluid mixing of the rhombic micromixers. Microfluid Nanofluid 4:419–425. https://doi.org/10.1007/s10404-007-0197-9
10. Afzal A, Kim KY (2015) Multi-objective optimization of a passive micromixer based on periodic variation of Velocity profile. Chem Eng Commun 202:322–333. https://doi.org/10.1080/00986445.2013.841150
11. Yakhshi-Tafti E, Kumar R, Cho HJ (2008) Effect of laminar velocity profile variation on mixing in microfluidic devices: the sigma micromixer. Appl Phys Lett 93:1–3. https://doi.org/10.1063/1.2996564
12. CFX-12.1 Solver Theory (2006) ANSYS Inc., Canonsburg, Pennsylvania
13. Cortes-Quiroz CA, Zangeneh M, Goto A (2009) On multi-objective optimization of geometry of staggered hertringbone micromixer. Microfluid Nanofluid 7:29–43. https://doi.org/10.1007/s10404-008-0355-8

14. Hossain S, Ansari MA, Kim KY (2009) Evaluation of the mixing performance of three passive micromixers. Chem Eng J 150:492–501. https://doi.org/10.1016/j.cej.2009.02.033
15. Yoshimura M, Shimoyama K, Misaka T, Obayasi S (2019) Optimization of passive grooved micromixers based on genetic algorithm and graph theory. Microfluid Nanofluid 23(30):1–21. https://doi.org/10.1007/s10404-019-2201-6
16. Orsi G, Roudgar M, Brunazzi E, Galletti C, Mauri R (2013) Water-ethanol mixing in T-shaped microdevices. Chem Eng Sci 95:174–183. https://doi.org/10.1016/j.ces.2013.03.015
17. Dizechi M, Marschall E (1982) Viscosity of some binary and ternary liquid mixtures. J Chem Eng Data 27(3):358–363. https://doi.org/10.1021/je00029a039
18. Wu C, Tang K, Gu B, Deng J, Liu Z, Wu Z (2016) Concentration-dependent viscous mixing in microfluidics: modelings and experiments. Microfluid Nanofluid 20(90):1–11. https://doi.org/10.1007/s10404-016-1755-9
19. Liu M (2011) Computational study of convective-diffusive mixing in a microchannel mixer. Chem Eng Sci 66:2211–2223
20. Chen H, Meiners JC (2004) Topologic mixing on a microfluidic chip. Appl Phys Lett 84(12):2193–2195. https://doi.org/10.1063/1.1686895
21. Patankar SV (1980) Numerical heat transfer and fluid flow. McGraw-Hill, New York
22. Ferziger JH, Peric M (2002) Computational methods for fluid dynamics. Springer, Berlin Heidelberg
23. Ansari MA, Kim KY, Anwar K, Kim SM (2010) A novel passive micromixer based on unbalanced splits and collisions of fluid streams. J Micromech Microeng 20:1–10. https://doi.org/10.1088/0960-1317/20/5/055007
24. Afzal A, Kim KY (2012) Passive split and recombination micromixer with convergent-divergent walls. Chem Eng J 203:182–192. https://doi.org/10.1016/j.cej.2012.06.111
25. Glatzel T, Litterst C, Cupelli C, Lindemann T, Moosmann C, Niekrawietz R, Streule W, Zengerle R, Koltay P (2009) Computational fluid dynamics (CFD) software tools for microfluidic applications—a case study. Comp Fluid 37:218–235. https://doi.org/10.1016/j.compfluid.2007.07.014
26. Jiang F, Drese KS, Hardt S, Küpper M, Schönfeld F (2004) Helical flows and chaotic mixing in curved micro channels. AIChE J 50:2297–2305. https://doi.org/10.1002/aic.10188
27. Bertsch A, Heimgartner S, Cousseau P, Renaud P (2001) Static mixers based on large-scale industrial mixer geometry. Lab Chip 1:56–60. https://doi.org/10.1039/B103848F
28. Wang H, Iovenitti P, Harvey E, Masood S (2003) Numerical investigation of mixing in microchannels with patterned grooves. J Micromech Microeng 13:801–808. https://doi.org/10.1088/0960-1317/13/6/302
29. Aubin J, Fletcher DF, Bertrand J, Xuereb C (2003) Characterization of the mixing quality in micromixers. Chem Eng Tech 26:1262–1270. https://doi.org/10.1002/ceat.200301848
30. Aubin J, Fletcher DF, Xuereb C (2005) Design of micromixers using CFD modeling. Chem Eng Sci 60:2503–2516. https://doi.org/10.1016/j.ces.2004.11.043
31. Kang TG, Kwon TH (2004) Colored particle tracking method for mixing analysis of chaotic micromixers. J Micromech Microeng 14:891–899
32. Afzal A, Kim KY (2014) Three-objective optimization of a staggered herringbone micromixer. Sens Actuators B: Chem 192:350–360. https://doi.org/10.1016/j.snb.2013.10.109
33. FLUENT-12.1, Solver Theory, ANSYS Inc., Canonsburg, Pennsylvania, 2006
34. Alam A, Afzal A, Kim KY (2014) Mixing Performance of a planar micromixer with circular obstructions in curved microchannel. Chem Eng Res Des 92:423–434. https://doi.org/10.1016/j.cherd.2013.09.008
35. Lee SW, Kim DS, Lee SS, Kwon TH (2006) A split and recombination micromixer fabricated in a PDMS three-dimensional structure. J Micromach Microeng 16:1067–1072. https://doi.org/10.1088/0960-1317/16/5/027
36. Lee SW, Lee SS (2008) Rotation effect in split and recombination micromixing. Sens Actuators B 129:364–371. https://doi.org/10.1016/j.snb.2007.08.038

37. Ansari MA, Kim KY (2010) Mixing performance of unbalanced split and recombine micromixers with circular and rhombic sub-channels. Chem Eng J 162:760–767. https://doi.org/10.1016/j.cej.2010.05.068

38. Afzal A, Kim KY (2015) Multi-objective Optimization of a Micromixer with Convergent-divergent sinusoidal walls. Chem Eng Commun 202:1324–1334. https://doi.org/10.1080/00986445.2014.935352

Chapter 4
Design Optimization of Micromixers

Abstract The mixing performance of a passive micromixer is sensitive to the geometry of the flow passages. Therefore, it is important to determine optimal configuration which maximizes the mixing performance of the micromixer. But, unfortunately, in some micromixers, enhancement of mixing performance is accompanied by a corresponding increase in pressure drop. Therefore, it is important to determine several configurations which represent the trade-offs between mixing efficiency and pressure drop. Numerical optimization techniques coupled with CFD analyses of flow and mixing have been proved to be an important tool for micromixer design. Both the single-objective and multi-objective optimization procedures for the shape optimization of micromixers are presented.

Keywords Micromixers · CFD · Design optimization · Surrogate modeling

4.1 Optimization Strategies for Micromixers

For passive micromixers, changes in the geometry can alter the flow and mixing dynamics inside the channel, and thus affect the performance of the devices. Many studies have been conducted to analyze the effects of geometrical and flow parameters on mixing performance leading to workable designs of micromixers. Using more sophisticated techniques such as design optimization [1], more efficient designs can be realized. There have been various approaches to design of micromixers, such as parametric study [2–7], layout optimization by solving a variational optimization problem [8], and design optimization using systematic optimization techniques [9–15].

Aubin et al. [3, 4] studied the effects of geometrical parameters on a SHM using a particle tracking approach and CFD. Three design parameters, viz. the depth and width of the grooves, and number of grooves per cycle, were tested. They quantified the mixing by analyzing the maximum striation thickness and residence time for various combinations of design parameters. Figure 4.1 shows the effect of groove width on mixing pattern at different axial locations. Qualitative information on the secondary flow inside the microchannel can be obtained from such plots. For narrow

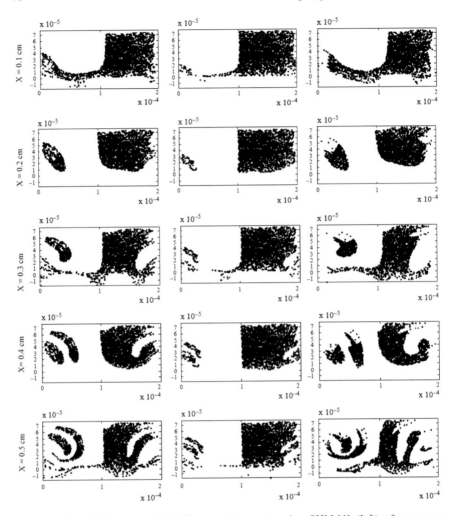

Fig. 4.1 Effect of the groove width, W_g, on mixing patterns in a SHM [4]: (left) reference case ($W_g = 50$ μm); (middle) $W_g = 25$ μm; (right) $W_g = 75$ μm

grooves, exchange of particles between the left-hand and right-hand sides of the channel is very poor, indicating a weak secondary flow. As the grooves become wider, the exchange of particles is enhanced. Overall, wide and deep grooves tend to promote mixing inside the SHM, while mixing is nearly unaffected by the number of grooves per cycle.

A similar study was conducted by Wang et al. [2] on a SHM. Using CFD simulation and particle tracking technique, Poincare maps were generated to study the chaotic flow in the SHM. Yang et al. [16] determined the effects of various geometrical parameters on mixing performance, flow rate, and pressure drop of a SHM using the Taguchi method and numerical simulation. Kang and Kwon [5] numerically investigated the mixing behavior of three different grooved micromixers, viz. slanted groove micromixer, SHM and barrier embedded micromixer. A 'colored particle tracking method' was developed to study the mixing performance both qualitatively and quantitatively.

Liu et al. [8] solved a variational optimization problem to obtain different types of passive micromixers using a layout optimization method. Constraints for the variational problem were Navier–Stokes equations and the convection–diffusion equation, with the mixing performance as an objective function. Figure 4.2 shows numerical simulation results for a passive micromixer designed by the layout optimization method, where measurements reveal a good mixing performance. They showed that the proposed optimization framework could reduce the dependency on the experience and intuition of designers. Experiments were conducted to support the effectiveness of the layout optimization method for conceptual design of the micromixer.

Numerical optimization techniques [9–15] based on CFD analysis have been proven to be an effective tool for robust and efficient design of passive micromixers. The objective function(s) for the optimization of a micromixer can be selected among the performance parameters such as mixing index, pressure loss, and residence time, etc. In the case of micromixers, the important performance parameters are the mixing efficiency and pressure loss. The mixing efficiency is the critical performance parameter related to the mixing performance of the device. The pressure loss is directly related to the pumping power required to drive the fluids through the micromixers. The optimization can be single-objective (e.g., to maximize the mixing efficiency) or multi-objective (e.g., to maximize the mixing efficiency and minimize the pressure loss).

Generally, the optimization procedure requires a large number of evaluations for the objective function(s) rendering conventional optimization techniques to be very expensive. To reduce the computational cost, surrogate model(s) is used to generate functional relationship between inputs and outputs, which is reliable representation of the simulation model as closely as possible. Queipo et al. [17] and Forrester and Keane [18] reviewed various surrogate models used in aerospace applications. Surrogate-based analysis and optimization have been applied to various optimization problems [19–24].

Fig. 4.2 Series-wound extension of the bending cells obtained by the layout optimization method [8]

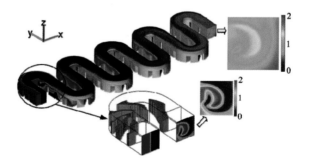

The SHM developed by Stroock et al. [25] has been used by many researchers to form a well-posed design optimization problem [9, 10, 13, 26]. As explained earlier in various studies [3, 4], the groove shape can significantly affect the mixing performance of SHMs. Ansari and Kim [9, 10] used radial basis neural network (RBNN) and response surface approximation (RSA) models for surrogate modeling in an optimization of a SHM. Afzal and Kim [14] performed an optimization of geometry and operating conditions of a convergent-divergent micromixer coupled with pulsatile flow to maximize the mixing performance. Using various surrogate models and sequential quadratic programming algorithm, an optimum design with a mixing index of 92.35% was obtained (Fig. 4.3).

Contrary to a single-objective optimization, a multi-objective optimization problem involves multiple conflicting design objectives. There are two approaches to a multi-objective optimization problem. In the first approach, multiple objectives can be combined using weights to form a single objective function. The estimation of the weights depends on the preference of the designer. Hossain et al. [12] conducted an optimization of a micromixer based on modified Tesla structure with weighted-average surrogate models. The objectives, i.e. the mixing index and pressure drop,

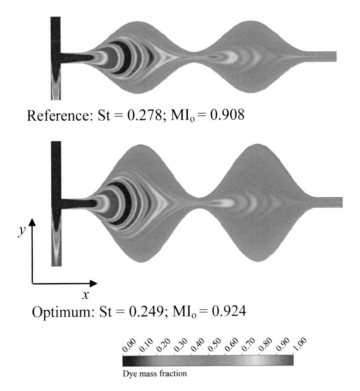

Reference: $St = 0.278$; $MI_o = 0.908$

Optimum: $St = 0.249$; $MI_o = 0.924$

Dye mass fraction

Fig. 4.3 Dye mass fraction distributions in the central x–y plane at $t = 0/T$ in reference and optimum designs [14]

were linearly combined using a weighting factor to yield a single-objective function. The other approach to multi-objective optimization uses multi-objective evolutionary algorithms (MOEAs), where multiple trade-off solutions for the objectives are determined. The latest MOEAs include Pareto evolutionary algorithms [27], Pareto archived evolutionary strategies [28] and an elitist non-dominated sorting genetic algorithm [29].

Multi-objective problems yield many solutions, which are known as Pareto-optimal solutions. These solutions can be used to analyze the trade-offs among designs. A multi-objective genetic algorithm (MOGA) was used by Hossain et al. [13] and Cortes-Quiroz et al. [26] for the shape optimizations of SHMs. The degree of mixing and the pressure drop were used as the objective functions. A Pareto-optimal front was established with an optimized trade-off between the maximum mixing index and the minimum pressure loss. In another study, Cortes-Quiroz et al. [11] carried out a multi-objective optimization of a passive micromixer with fin-shaped baffles in a T-channel to obtain Pareto-optimal designs.

Afzal and Kim carried out multi-objective optimizations of a Sigma micromixer [30], an SAR micromixer with convergent-divergent sinusoidal walls [31] and a SHM [32]. Surrogate models, viz. RSA and RBNN were used to approximate the objective functions: mixing index and non-dimensional pressure drop. The surrogate models for the objectives were supplied as fitness functions to MOGA to obtain the Pareto-optimal front. Figure 4.4 shows a Pareto-optimal front representing the trade-off between conflicting objectives, mixing index and pressure drop [30]. A designer

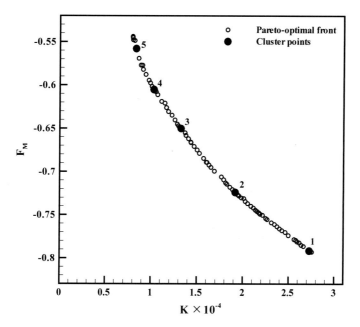

Fig. 4.4 Pareto-optimal front representation of the non-dominated solutions between two objectives [30]

may pick a solution on the Pareto-optimal front depending on his/her needs. Afzal and Kim [31] used mixing effectiveness to pick an optimum design among available Pareto-optimal solutions.

4.2   Conventional Versus Surrogate-Based Optimization

Various approaches have been used by researchers for optimization of a typical system characterized by a set of inputs and responses. The most basic approach is parametric study, which involves studying the effects of each design parameter on system responses, keeping other parameters fixed. A more advanced approach is the use of design of experiment (DOE) techniques. In DOE procedures, multiple parameters can be manipulated determining their effects on system response. Examples of DOE are factorial designs, Latin hypercube sampling, etc.

Using the above mentioned approaches, it is possible to find acceptable or workable designs. Nonetheless, the urge to find the optimum design remains the most important part. Therefore, optimization can be performed to search the entire design space using a suitable optimization algorithm to maximize or minimize a system response. It can be achieved using either gradient–based methods for constrained and unconstrained optimization, or heuristic techniques like particle swarm optimization.

Having understood the importance of optimization, Figs. 4.5 and 4.6 show two different strategies for optimization, which are conventional and surrogate-based optimizations, respectively. In order to evaluate the objective function(s) in optimizations of fluid and thermal systems, the key step is to analyze the phenomena using Navier–Stokes equations coupled with transport equation for heat/mass transfer together with appropriate boundary conditions. This step is inherent to either of the optimization strategies. The difference lies in the coupling of numerical model with the optimization algorithm. In conventional optimization, the numerical model is coupled directly with the optimization algorithm, and the numerical model needs to

Fig. 4.5 Conventional optimization (an example)

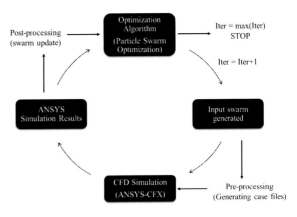

Fig. 4.6 Surrogate-based
optimization (an example)

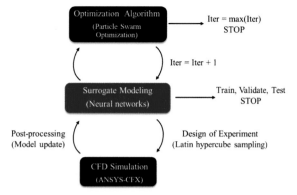

be run for every design outcome of the optimization algorithm. Because of this, one
of the major drawback of this method is that the computational cost is prohibitively
large for entire design optimization process.

In this respect, surrogate modeling has been used to reduce the computational
burden with a reliable representation of the simulation data. Using CFD simulation
results, a final surrogate model is approved using a suitable approach which involves
training, testing and validation in the case of using neural networks. This approved
model is supplied as fitness function to the optimization algorithm which yields the
optimum design in a fast and efficient way. To get the best results using this strategy,
the efficacy of the surrogate model is evaluated in terms of global exploration and
local exploitation characteristics before it is approved for coupling with the optimizer.
Surrogate-based optimizations have been extensively used for design optimization
of micromixers.

4.3 Design of Experiments

Design of Experiments (DOE) procedures are used to sample the design variable
space. It is conducted to determine the relationship between the different factors
affecting a process and the output of that process to extract maximum amount of
information. For optimization, DOE methods are used to generate design sites to
build a surrogate model. In particular, a good sampling plan can be efficiently used
for fitting a variety of models. One of the most commonly used DOE methods is
Latin hypercube sampling (LHS) which has a form of stratified sampling that can be
applied to multiple variables [33, 34]. Using McKay et al. notation [33], a sample of
size N can be constructed by dividing the range of each factor (input variable) into N
strata of equal marginal probability $1/N$ and sampling once from each stratum. Also,
the uniformity of the sampling plan can be controlled using uniformity measures
like maximum-minimum distance between the points, or by correlation among the
sample data.

4.4   Surrogate Modeling

In optimization of thermo-fluid problems, a large number of numerical analyses are required to evaluate objective function(s), but a single numerical simulation takes long time to complete due to the non-linear governing differential equations. Surrogate modeling is introduced to alleviate this burden by constructing approximation models, which mimic the behavior of the simulation model as closely as possible. The major advantages of surrogate modeling are computational economy in evaluating the objective function(s), and accuracy in representing the characteristics of the design space. The mathematical formulations for different surrogate models are discussed below.

4.4.1   Response Surface Approximation

Response surface approximation (RSA) model is the simplest, yet widely used surrogate model for studying the underlying relationship in the data [35]. RSA models are extension of linear regression models which contain additional features/predictors generated using non-linear transformation of the input space to account for non-linearity in the model.

In the linear regression model, a continuous response $\mathbf{y}$ is usually modeled as a linear combination of the predictor, $\mathbf{x} = \begin{bmatrix} x_1 \ x_2 \ \ldots \ x_p \end{bmatrix}^T$ plus a random error, $\in$. For N independent observations (i $= 1\ldots$, N) of p design variables of the predictor $\mathbf{x}$ and response $\mathbf{y}$, the model can be represented as:

$$y_i = \beta_0 + \beta_1 x_{1,i} + \beta_2 x_{2,i} + \cdots + \beta_p x_{p,i} + \in_i (0, \sigma_\in^2) \tag{4.1}$$

where β_j (j $= 0\ldots$, p) are the coefficients of the model. Errors, $\in_i$ are assumed to be uncorrelated and distributed with a mean of 0 and constant variance, $\sigma_\in^2$. In compact form, Eq. (4.1) can be written as:

$$\mathbf{Y} = \mathbf{X}\boldsymbol{\beta} + \in \tag{4.2}$$

$$\mathbf{X} = \begin{bmatrix} 1 & x_{1,1} & x_{2,1} & \cdots & x_{p,1} \\ \vdots & \vdots & \vdots & \cdots & \vdots \\ 1 & x_{1,i} & x_{2,i} & \cdots & x_{p,i} \\ \vdots & \vdots & \vdots & \cdots & \vdots \\ 1 & x_{1,N} & x_{1,N} & \cdots & x_{p,N} \end{bmatrix}, \mathbf{Y} = \begin{bmatrix} y_1 \\ \vdots \\ y_i \\ \vdots \\ y_N \end{bmatrix}$$

$\mathbf{X}$ is the design matrix of the system. For RSA model, Eq. (4.1) can be modified as:

$$y_i = \sum_{j=0}^{m} \beta_j \varphi_j^{(i)} + \epsilon_i \ (0, \sigma_\epsilon^2) \tag{4.3}$$

where φ_j are basis-functions. The design matrix, $\mathbf{X}$ now contains additional terms obtained using non-linear transformation of the input variables for sampled points using the basis-functions. For model approximation, the coefficients β_j ($j = 0..., p$) need to be determined. Using sample data, the least-square estimate of β is:

$$\beta = \mathbf{X}^\dagger \mathbf{Y} \tag{4.4}$$

$$\mathbf{X}^\dagger = (\mathbf{X}^T \mathbf{X})^{-1} \mathbf{X}^T \tag{4.5}$$

$\mathbf{X}^\dagger$ is known as pseudo-inverse of $\mathbf{X}$. The choice of basis-functions will affect the accuracy of RSA model, and it becomes important to determine the right basis-functions φ_j as well as associated model coefficients β_j to obtain the best model on the sample data. To determine the best model which approximates the data, the goodness of fit, R_{adj}^2 is estimated. For a good fit, R_{adj}^2 must be closer to 1.

4.4.2 Radial Basis Neural Networks

Radial basis neural networks (RBNN) is a two-layered network consisting of a hidden layer of radial basis neurons and an output layer of linear neurons, characterized by a set of inputs and a set of outputs [36]. The radial basis functions act as processing units between the input and output. The hidden layer performs a non-linear transformation of the input space to an intermediate space using a set of radial basis units.

A Gaussian function of the following form is used as a transfer function for a radial basis neuron.

$$\phi(\mathbf{x}) = \exp\left(-\gamma \|\mathbf{x} - \mathbf{c}_i\|^2\right) \tag{4.6}$$

The net input to the transfer function is the vector distance between the neurons center, $\mathbf{c}_i$ and the input vector, $\mathbf{x}$ multiplied by the parameter, γ. The parameter γ allows the sensitivity of the neurons to be adjusted. The output layer, then implements a linear combiner to produce the desired targets. The output function can be expressed as:

$$h(\mathbf{x}) = \sum_{i=1}^{K} w_i \exp\left(-\gamma \|\mathbf{x} - \mathbf{c}_i\|^2\right) \tag{4.7}$$

The prediction ability of the network is stored in the weights, w_i which can be obtained from a set of training data [36].

4.4.3 Kriging

The Kriging (KRG) model [37] can be formulated as a combination of two components, global model and a systematic departure,

$$y(x) = \overline{y}(x) + Z(x) \tag{4.8}$$

where $y(x)$ is the unknown function to be estimated and $\overline{y}(x)$ is a known function (usually a regression function) representing the trend over the design space, also called the global model. The second term, $Z(x)$ creates a localized deviation to interpolate the sampled data points by quantifying the correlation of the points with a Gaussian correlation having a zero mean and nonzero covariance.

The three main Kriging variants, simple, ordinary and universal, differ in their treatment of the trend component, μ. The most popular universal kriging model can be expressed as:

$$y(x) = \sum_{j=1}^{n} \alpha_j g_j(x) + Z(x) \tag{4.9}$$

where $g_j(x)$ are regressors, and α_j are model coefficients in linear regression. The covariance function is generally derived from the input semivariogram model.

The Gaussian function is the most commonly used due to superior numerical properties like providing an infinitely differentiable surface and easy integration with gradient-based optimization algorithms. The mathematical form of the univariate Gaussian function is:

$$cov(x_1, x_2) = \sigma_f^2 R(x_1, x_2) \tag{4.10}$$

$$R(x_1 - x_2) = e^{-\theta|x_1 - x_2|^2}, \theta > 0 \tag{4.11}$$

The variance, σ_f^2 scales the spatial correlation function, $R(x_1, x_2)$, and parameter, θ controls the width of the Gaussian function. For multivariate correlation function, a univariate correlation function is used for each of the d input dimensions, and a product correlation rule is used:

$$R(x_1, x_2) = \prod_{i=1}^{d} R(|x_{2,i} - x_{1,i}|) \tag{4.12}$$

The parameters of the Kriging model, $\left\{\alpha, \sigma_f^2, \theta\right\}$ are found by using the maximum likelihood estimation (MLE). More details on the Kriging model, and its implementation can be found in the references [14, 37, 38].

4.5 Single-Objective Optimization

The standard form of a single-objective optimization problem is:

$$\underset{x}{maximize} \, / \, \underset{x}{minimize} \, f(x)$$

$$subject \, to \, x_{min} \leq x \leq x_{max}$$

where $f(x)$ is an objective function and x is a design variable vector. x_{min} and x_{max} are vectors for the lower and upper bounds of the design variables, respectively.

In the case of micromixers, mixing index is a critical performance parameter related to the performance of the device. In most cases [9, 10, 12, 14], the mixing index evaluated using the method described in Sect. 3.2 was used as the objective function, $f(x)$. Parametric analysis is usually conducted to select the design variables and their ranges for optimization. The detailed flowchart for the optimization procedure is shown in Fig. 4.7.

The design space bounded by the lower and upper bounds of the chosen design variables are discretized using DOE such as LHS to generate design points, which are further used to construct the surrogate models to approximate the objective function. CFD simulations are carried out to determine the objective function values at the design points. A surrogate model of the objective function, $h(x)$ is approved using a suitable error estimation technique, and is supplied as fitness function to the optimization algorithm to determine the optimum point. Once the optimum point is found by an algorithm, it is verified using CFD analysis. Some of the algorithms used to find the optimum point on the surrogate model are: sequential quadratic programming (SQP) [39], particle swarm optimization (PSO) [40, 41], genetic algorithm (GA) [42], simulated annealing (SA) [43], among others.

Afzal and Kim [4] performed a comparative evaluation of various global optimization algorithms for three practical design applications in the field of thermo-fluids engineering. Figure 4.8 shows the results of this comparative analysis for optimization of a convergent-divergent micromixer coupled with pulsatile flow. Four different algorithms were tested: GA, PSO, SA, and SQP, and their performances were evaluated comparatively. Among the tested algorithms, PSO showed the best overall performance in the combined aspects of optimization result and computational time.

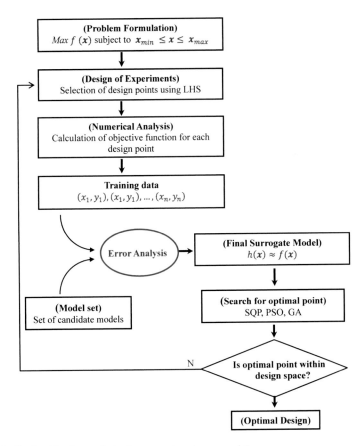

Fig. 4.7 Single-objective optimization procedure (an example)

4.6 Multi-Objective Optimization

The standard form of a multi-objective optimization problem is:

$$\underset{x}{maximize} \,/\, \underset{x}{minimize} \, \boldsymbol{f}(\boldsymbol{x})$$

$$subject\ to\ \boldsymbol{x}_{min} \leq \boldsymbol{x} \leq \boldsymbol{x}_{max}$$

where $\boldsymbol{f}(\boldsymbol{x}) = \left[f_1(\boldsymbol{x}), f_2(\boldsymbol{x}), \ldots, f_p(\boldsymbol{x}) \right]$ is a vector of objective functions and $\boldsymbol{x}$ is a vector of design variables. $\boldsymbol{x}_{min}$ and $\boldsymbol{x}_{max}$ are vectors for the lower and upper bounds of the design variables, respectively.

A multi-objective problem yields many solutions, which are known as Pareto-optimal solutions. Each feasible solution set $\boldsymbol{x}$ of the multi-objective optimization

Fig. 4.8 Optimum objective
function values calculated by
CFD (top), prediction errors
(middle), and computational
times for different trials
(bottom) for a
convergent-divergent
microchannel coupled with
pulsatile flow [44]

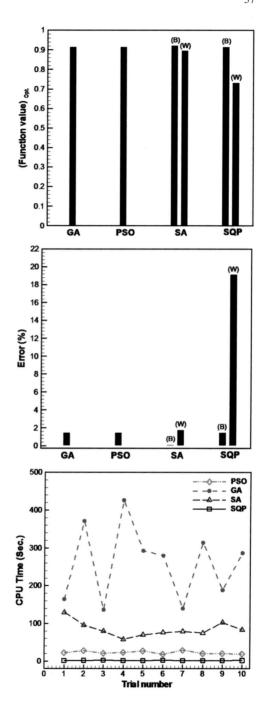

problem is either dominated or non-dominated. The relative dominance can be established using the following conditions: A design, x_1 dominates a design, x_2, if x_1 is at least as good as x_2 for all objectives and x_1 is strictly better than x_2 for at least one objective. All of the designs that are non-dominated to any other design, comprise a Pareto-optimal set. The functional space representation of the Pareto-optimal solution set is the Pareto-optimal front. The Pareto-optimal solution represents the trade-off among conflicting objectives, and can be used to analyze the trade-offs among designs. Because each solution is a global Pareto-optimal solution, none of these Pareto-optimal solutions is superior to the others for both objectives. Thus, the choice by the designer is important when selecting a Pareto-optimal solution that meets a given requirement.

To obtain Pareto-optimal solutions, a MATLAB built-in function, *gamultiobj*, can be used to invoke MOGA [30–32]. The function uses a controlled elitist genetic algorithm (a variant of NSGA-II [29]) which uses elite individuals differently than the genetic algorithm. It sorts non-inferior individuals above inferior ones, thus the elite individuals are automatically used. In non-elitist MOGAs, the genetic operator may destroy some of the non-dominated solutions to explore the design space. Introducing elitism in MOGAs alleviates this problem to some extent [45]. A hybrid function was used for subsequent minimization after the genetic algorithm was terminated. The hybrid solver starts at all points on the Pareto front returned by MOGA. The new individuals returned by the hybrid solver are combined with the existing population, and a new Pareto front is obtained. The algorithm terminates based on the convergence criterion specified by the user. The inputs to the algorithm can affect the development of Pareto-optimal front, and therefore, several cases need to be simulated to select the correct parameters for the genetic algorithm.

The optimization procedure to obtain the Pareto-optimal solutions is shown in Fig. 4.9. As mentioned earlier in Sect. 3.2, the mixing index is the primary objective function related to the performance of a micromixer. Another candidate for objective function is pressure loss. In the multi-objective optimizations of Afzal and Kim [30, 31], the aim was to simultaneously maximize mixing performance and minimize pressure loss in a multi-objective optimization framework.

A Pareto-optimal front is established with optimized trade-offs among different objective functions. The obtained Pareto-optimal front and the corresponding Pareto-optimal designs (PODs) need to be analyzed to identify relations among different objectives. Therefore, some of representative PODs, which cover both the design and functional spaces, are selected by performing *k-means* clustering. It is an iterative process for forming clusters [46]. Further, numerical simulations are conducted to determine the objective function values at the representative PODs obtained from clustering. These values are then compared with the corresponding objective function values obtained from the optimization. This exercise helps to determine the relative accuracies of the CFD modeling, adopted function approximations, and multi-objective optimization algorithm.

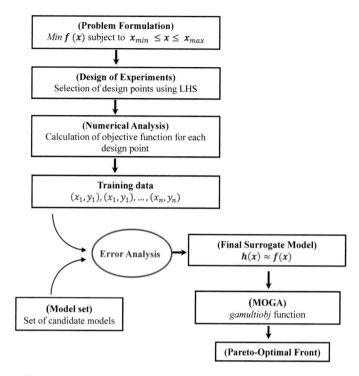

Fig. 4.9 Multi-objective optimization procedure (an example)

References

1. Kim KY, Samad A, Benini E (2019) Design optimization of fluid machinery (Applying Computational Fluid Dynamics and Numerical Optimization). Wiley, Singapore
2. Wang H, Iovenitti P, Harvey E, Masood S (2003) Numerical investigation of mixing in microchannels with patterned grooves. J Micromech Microeng 13:801–808. https://doi.org/10.1088/0960-1317/13/6/302
3. Aubin J, Fletcher DF, Bertrand J, Xuereb C (2003) Characterization of the mixing quality in micromixers. Chem Eng Tech 26:1262–1270. https://doi.org/10.1002/ceat.200301848
4. Aubin J, Fletcher DF, Xuereb C (2005) Design of micromixers using CFD modeling. Chem Eng Sci 60:2503–2516. https://doi.org/10.1016/j.ces.2004.11.043
5. Kang TG, Kwon TH (2004) Colored particle tracking method for mixing analysis of chaotic micromixers. J Micromech Microeng 14:891–899
6. Shakhawat H, Kim KY (2010) Numerical study on mixing performance in straight groove micromixers. Int J Fluids Mach And Sys 3:227–234. https://doi.org/10.5293/IJFMS.2010.3.3.227
7. Afzal A, Kim KY (2014) Performance Evaluation of three types of passive micromixer with Convergent-divergent sinusoidal walls. J Mar Sci Tech –Taiwan 22(6):680–686. https://doi.org/10.6119/JMST-014-0321-2
8. Liu Y, Deng Y, Zhang P, Liu Z, Wu Y (2013) Experimental investigation of passive micromixers conceptual design using layout optimization method. J Micromech Microeng 23:1–10. https://doi.org/10.1088/0960-1317/23/7/075002

9. Ansari MA, Kim KY (2007) Shape optimization of a micromixer with staggered herringbone groove. Chem Eng Sci 62:6687–6695. https://doi.org/10.1016/j.ces.2007.07.059
10. Ansari MA, Kim KY (2007b) Application of the radial basis neural network to optimization of a micromixer. Chem Eng Tech 30:962–966. https://doi.org/10.1002/ceat.200700055
11. Cortes-Quiroz CA, Azarbadegan A, Moeendarbary E (2010) An efficient passive planar micromixer with fin-shaped baffles in the tee channel for wide Reynolds number flow range. World Acad Sci Eng Technol 61:170–175. https://doi.org/10.5281/zenodo.1333420
12. Hossain S, Ansari MA, Husain A, Kim KY (2010) Analysis and optimization of a micromixer with a modified Tesla structure. Chem Eng J 158:305–314. https://doi.org/10.1016/j.cej.2010.02.002
13. Hossain S, Husain A, Kim KY (2011) Optimization of micromixer with staggered herringbone grooves on top and bottom walls. Eng Appl Comput Fluid Mech 5:506–516. https://doi.org/10.1080/19942060.2011.11015390
14. Afzal A, Kim KY (2015) Optimization of pulsatile flow and geometry for a Convergent-divergent micromixer. Chem Eng J 281:134–143. https://doi.org/10.1016/j.cej.2015.06.046
15. Hossain S, Afzal A, Kim KY (2017) Shape optimization of a three-dimensional serpentine split-and- recombine Micromixer. Chem Eng Commun 204(5):548–556. https://doi.org/10.1080/00986445.2017.1289185
16. Yang JT, Huang KJ, Lin YC (2005) Geometric effects on fluid mixing in passive grooved micromixers. Lab Chip 5:1140–1147. https://doi.org/10.1039/B500972C
17. Queipo NV, Haftka RT, Shyy W, Goel T, Vaidyanathan R, Tucker PK (2005) Surrogate-based analysis and optimization. Prog Aerosp Sci 41:1–28. https://doi.org/10.1016/j.paerosci.2005.02.001
18. Forrester AI, Keane AJ (2009) Recent advances in surrogate-based optimization. Prog Aerosp Sci 45:50–79. https://doi.org/10.1016/j.paerosci.2008.11.001
19. Kim HM, Kim KY (2006) Shape optimization of three-dimensional channel roughened by angled ribs with RANS analysis of turbulent heat transfer. Int J Heat Mass Transf 49:4013–4022. https://doi.org/10.1016/j.ijheatmasstransfer.2006.03.039
20. Kim KY, Seo SJ (2006) Application of numerical optimization technique to design of forward-curved blades centrifugal fan. JSME Int J Ser B 49:152–158. https://doi.org/10.1299/jsmeb.49.152
21. Kulkarni K, Afzal A, Kim KY (2015) Multi-objective optimization of Solar Air heater with Obstacles on Absorber Plate. Sol Energy 114:364–377. https://doi.org/10.1016/j.solener.2015.02.008
22. Seo J-W, Afzal A, Kim KY (2016) Efficient multi-objective optimization of a boot-shaped rib in a cooling channel. Int J Therm Sci 106:122–133. https://doi.org/10.1016/j.ijthermalsci.2016.03.015
23. Kulkarni K, Afzal A, Kim KY (2016) Multi-objective optimization of a double-layered microchannel heat sink with temperature-dependent fluid properties. Appl Therm Eng 99:262–272. https://doi.org/10.1016/j.applthermaleng.2016.01.039
24. Kim SM, Afzal A, Kim KY (2016) Optimization of a Staggered Jet-Convex Dimple Array Cooling System. Int J Therm Sci 99:161–169. https://doi.org/10.1016/j.ijthermalsci.2015.08.013
25. Stroock AD, Dertinger SK, Ajdari A, Mezić I, Stone HA, Whitesides GM (2002) Chaotic mixer for microchannels. Science 295:647–651. https://doi.org/10.1126/science.1066238
26. Cortes-Quiroz CA, Zangeneh M, Goto A (2009) On multi-objective optimization of geometry of staggered hertringbone micromixer. Microfluid Nanofluid 7:29–43. https://doi.org/10.1007/s10404-008-0355-8
27. Zitzler E, Thiele L (1998) An evolutionary algorithm for multiobjective optimization: The strength Pareto approach. TIK-report 43. https://doi.org/10.3929/ethz-a-004288833
28. Knowles JD, Corne DW (2000) Approximating the non-dominated front using the Pareto archived evolution strategy. Evol Comput 8:149–172. https://doi.org/10.1162/106365600568167
29. Deb K (2001) Multi-objective optimization using evolutionary algorithms. Wiley

30. Afzal A, Kim KY (2015) Multi-objective Optimization of a Passive Micromixer based on periodic variation of Velocity profile. Chem Eng Commun 202:322–333. https://doi.org/10.1080/00986445.2013.841150
31. Afzal A, Kim KY (2014) Three-objective optimization of a staggered herringbone micromixer. Sens Actuat B: Chem 192:350–360. https://doi.org/10.1016/j.snb.2013.10.109
32. Afzal A, Kim KY (2015) Multi-objective Optimization of a Micromixer with Convergent-divergent sinusoidal walls. Chem Eng Commun 202:1324–1334. https://doi.org/10.1080/00986445.2014.935352
33. McKay MD, Beckman RJ, Conover WJ (1979) A comparison of three methods for selecting values of input variables in the analysis of output from a computer code. Technometrics 21:239–245. https://doi.org/10.1080/00401706.1979.10489755
34. Stein M (1987) Large sample properties of simulations using latin hypercube sampling. Technometrics 29:143–151. https://doi.org/10.1080/00401706.1987.10488205
35. Myers RH, Montgomery DC, Anderson-Cook CM (1995) Response surface methodology: process and product optimization using designed experiments. Wiley, New York, pp 134–174
36. Chen S, Cowan CF, Grant PM (1991) Orthogonal least squares learning algorithm for radial basis function networks. IEEE Trans Neural Netw 2:302–309. https://doi.org/10.1109/72.80341
37. Martin JD, Simpson TW (2005) Use of Kriging models to approximate deterministic computer models. AIAA J 43(4):853–863
38. Raza W, Kim KY (2008) Shape optimization of wire—wrapped fuel assembly using kriging metamodeling technique. Nucl Eng Design 238(6):1332–1341. https://doi.org/10.1016/j.nucengdes.2007.10.018
39. Boggs PT, Tolle JW (2000) Sequential quadratic programming for large-scale nonlinear optimization. J Comp App Math 124:123–137. https://doi.org/10.1016/S0377-0427(00)00429-5
40. Kennedy J, Eberhart RC (1995) Particle swarm optimization. Proceedings of the IEEE international conference on neural networks, Perth, Australia
41. Holland JH (1975) Adaptation in natural and artificial systems. The University of Michigan Press, Ann Arbor
42. Goldberg DE (1989) Genetic algorithms in search, optimization, and machine learning. Addison-Wesley, Boston
43. Kirkpatrick S, Gelatt CD, Vecchi MP (1983) Optimization by simulated annealing. Science 220:671–680. https://doi.org/10.1126/science.220.4598.671
44. Ma SB, Afzal A, Kim KY (2018) Optimization of ring cavity in a centrifugal compressor based on comparative analysis of optimization algorithms. App Th Eng 99:262–272. https://doi.org/10.1016/j.applthermaleng.2018.04.094
45. Goel T, Vaidyanathan R, Haftka RT, Shyy W, Queipo NV, Tucker K (2007) Response surface optimization of Pareto-optimal front in multi-objective optimization. Comput Methods Appl Mech Eng 196:879–893. https://doi.org/10.1016/j.cma.2006.07.010
46. Lloyd SP (1982) Least squares quantization in PCM. IEEE Trans Inf Theory 28:129–137. https://doi.org/10.1109/TIT.1982.1056489

Chapter 5
Conclusion

Abstract This chapter summarizes the details of what have been covered in the previous chapters, and how to pursue those researches further.

Keywords Summary · Conclusions · Future Researches

The book is dedicated to analysis and design optimization of micromixers. The contents of the book were presented in a sequence starting from applications (Chap. 1), mixing concepts (Chaps. 1 and 2), flow and mixing analyses (Chap. 3) and design optimization (Chap. 4) of micromixers. Major emphases was laid on numerical analyses of flow and mixing in micromixers, and design optimization techniques.

Chapter 1 provides introduction to micromixers in light of wide variety of chemical and biological applications, types of micromixers, and characterization of flow and mixing regimes based on dimensionless numbers. A few selected applications of micromixing technology were presented, but the readers can refer to the cited references for more detailed and involved applications to develop a broad background on the researches. The dimensionless numbers were introduced to provide measures of relative effects which play important role in determining how mixing occurs in various micromixers. The analysis based on the dimensionless numbers can be of great help to practitioners and designers in classifying different micromixers, and determining suitability of a micromixer to a particular application.

Chapter 2 gives overview of different types of active and passive micromixers. Active micromixers uses external energy to perturb the flow field for efficient mixing. Some basic types of active micromixers were covered based on pressure field, pulsation, electrokinetics, magnetohydrodynamics, acoustics, and Coriolis induced flows. On the other hand, passive micromixers rely on the geometry of the devices to produce complex flow fields for mixing. Due to certain advantages of passive micromixing concept over the active type, different concepts and designs of passive micromixers were the focus of this chapter.

Two basic concepts were presented: multi-lamination and focusing which rely solely on molecular diffusion for mixing enhancement, and chaotic advection which

© The Author(s), under exclusive license to Springer Nature Singapore Pte Ltd. 2021   63
A. Afzal and K.-Y. Kim, *Analysis and Design Optimization of Micromixers*,
SpringerBriefs in Computational Mechanics,
https://doi.org/10.1007/978-981-33-4291-0_5

depends on stretching and folding of fluid streams for efficient mixing. Chaotic advection can be generated in a variety of ways, and the geometrical modifications using surface patterning, serpentine channels, obstacles on channel walls, split and recombination, and three-dimensional channel structures, were discussed. Due to limited length of the book, it was not possible to cover all active and passive micromixers, but some review papers focussed on active and passive micromixers were cited for the readers who wish to look into more detailed comparison of micromixers.

Mixing in micromixers needs to be quantified for the evaluation of performance, and the derived performance matrices can be used for the purpose of design optimization. Different numerical techniques to find the solutions of the governing differential equations for flow dynamics and mixing in micromixers were introduced in Chap. 3. The solutions for the velocity and concentration fields can be manipulated to derive performance measures. Since flow and mixing analyses are based on both the Eulerian and Lagrangian approaches, both the approaches were discussed separately in detail for better understanding and clarity to the readers. An overview on commercial CFD packages relevant to micromixers were presented. The challenges and limitations of numerical schemes, computational grid requirements, and the associated errors in the numerical solutions were also discussed. Chapter 3 concluded with the technique for mixing characterization using velocity and concentration fields, and forms the basis for performance evaluation of different micromixer designs. The chapter not only provides the fundamentals of numerical techniques, but also discusses their relative advantages and disadvantages for different types of mixing problems encountered in practice. Also, details of numerical models in popular commercial CFD packages, viz. CFD-ACE+®, ANSYS-CFX®, ANSYS-Fluent®, were briefed for end users who wish to conduct research on micromixers.

Chapter 4 was dedicated to design optimization of micromixers. Several optimization techniques were presented, but the major emphasis was laid on surrogate-based optimization as it was found to be the most widely used method for design optimization of micromixers. In case of micromixers, the optimization problems can be classified as either single-objective (e.g., to maximize mixing performance of the micromixer) or multi-objective (e.g., to maximize mixing performance and minimize pressure drop in the micromixer, simultaneously) problem. Conventional optimizations generally employ gradient-based methods which tend to get trapped in local optima rather than finding the global optimum solution, or population based methods like genetic algorithm which require impractical computational cost to obtain global solutions. Since simulation models used for design of micromixers are generally expensive, surrogate models are good alternative to reduce the computational burden with acceptable approximation of the simulation data. Surrogate model(s) for the objective function(s) coupled with an optimization algorithm can be used to find optimal geometric configuration(s) subjected to a set of design constraints. This chapter focused on the algorithm for single-objective and multi-objective optimizations using surrogate models for the design and development of micromixers.

Some suggestions for further researches on passive micromixers are: (1) one of the major goal in micromixing is to develop efficient micromixers, and therefore, novel micromixer designs need to be developed to achieve higher mixing performance over a shorter micromixer length, (2) a significant portion of the studies on micromixers assumed fluids to be Newtonian. However, in many practical microfluidic applications, the fluids are non-Newtonian such as bio-fluids, blood samples, and polymer solutions. Proper models for non-Newtonian fluid need to be incorporated in governing equations, and analysis can be done with respect to the different operating conditions encountered in microfluidic devices, (3) affordable computational grid is key requirement for mixing analysis in micromixers, and to reduce discretization error (i.e., numerical diffusion), numerical schemes for advection terms need to be further studied, (4) although optimizations of micromixer designs have been successfully performed by many researchers, the performance parameters were limited to mixing index and pressure drop. It would be interesting to test other performance parameters such as mixing time and energy dissipation. Also, the multi-objective optimization method can be extended to handle three- or more objectives, which would be useful for more robust design of microfluidic systems.

Printed in the United States
By Bookmasters